An Introduction to
Confirmation Theory

An Introduction to Confirmation Theory

RICHARD SWINBURNE

Professor of Philosophy,
University of Keele

METHUEN & CO LTD
11 NEW FETTER LANE LONDON EC4

First published 1973
by Methuen & Co Ltd
11 New Fetter Lane London EC4P 4EE
© 1973 R. Swinburne
Typesetting by William Clowes & Sons Ltd
London, Colchester and Beccles

Printed in Great Britain by
Lowe & Brydone (Printers) Ltd

SBN 416 66970 0

Distributed in the U.S.A. by
HARPER & ROW PUBLISHERS INC.
BARNES & NOBLE IMPORT DIVISION

Contents

434254

Preface

Confirmation Theory is the theory of when and how much different evidence renders different hypotheses probable. The aim of this book is to expound and criticize the views of philosophers on Confirmation Theory, and in the process to contribute towards the construction of a correct Confirmation Theory. I have placed at the end of each chapter a bibliography of the most important works discussed in the chapter, and these works are referred to by numbers in square brackets. In references to articles in journals the first number indicates the year of publication of the journal, the second number the volume of the journal, and the third number the pages of that volume which contain the article.

Some of the material has appeared in articles in journals, although some of the conclusions which I reach in this book are substantially different from conclusions reached in the articles. I am grateful to the editors of the following journals for permission to use material from the articles cited: *American Philosophical Quarterly* ('Probability, Credibility, and Acceptability', 1971, **8**, 275-83; 'The Paradoxes of Confirmation – A Survey', 1971, **8**, 318-30), *Analysis* ('Grue', 1968, **28**, 123-8; 'Projectible Predicates', 1969, **30**, 1-11), *Australasian Journal of Philosophy* ('Popper's Account of Acceptability', 1971, **49**, 167-76), *Mind* ('Cohen on Evidential Support', 1972, **81**, 244-8) and *Philosophy of Science* ('The Probability of Particular Events', 1971, **38**, 327-43).

I am very grateful to the many philosophers who criticized earlier versions of individual chapters; to Jonathan Cohen, Hugh Mellor and Alan White who provided helpful criticisms of the penultimate version of the book; and to my wife who corrected the final typescript and the proofs.

November 1971

CHAPTER I

Epistemic Probability

Often a proposition makes a second proposition probable or more probable than some other proposition, or confers on it a definite amount of probability, in the sense that to the extent stated it provides evidence which makes the second proposition likely to be true. That planets had been observed in certain particular positions made it probable that the planets describe ellipses with the sun at one focus. That Jones has defeated Smith in all the races which they have had so far makes it probable that Jones will defeat Smith in their next race. That the pack of cards in front of me is a well shuffled pack of fifty-two cards of which thirteen are hearts makes the probability that the top card is a heart 1/4. Confirmation Theory is concerned to state the rules for one proposition conferring probability on a second proposition in the above sense. The first proposition is often called the evidence, and the second the hypothesis. Confirmation Theory seeks to state the rules for different evidence conferring probability on different hypotheses. Confirmation Theory is not concerned so much with which hypotheses are currently made probable by present evidence, but with what kinds of characteristics evidence and hypothesis have to have in order that the evidence makes probable the hypothesis. Our evidence is what forms our starting point for inference, what we can be said to know. (I shall not discuss how we acquire new knowledge nor how we realize that we do not know some of the things which we thought we did.)

A similar relation to the one analysed above as holding between propositions, holds between the events or states of affairs reported by the propositions. In so far as 's occurs' makes probable 's¹ occurs', in the sense that the former provides evidence which makes the latter likely to be true, then the occurrence of s makes the occurrence of s^1 probable, in the sense that the occurrence of s provides evidence for the occurrence of s^1. Conversely if the occurrence of s makes the occurrence of s^1 probable in the stated sense, then 's occurs' makes probable in that sense 's^1 occurs'. These relations, whether between propositions or states of affairs, I will term relations of epistemic probability. I shall treat mainly of the epistemic probability which holds between propositions, and elucidate the criteria for one proposition making another probable. In virtue of the logical equivalence just stated, I shall thereby also be elucidating the criteria for one event making another event epistemically probable. Sometimes indeed I shall change to talking of events making other events probable.

Some writers who have attempted to elucidate the criteria of epistemic probability have treated it as holding between events; some have treated it as holding between propositions. Either method is possible for it holds between both and the relations between the two kinds of epistemic probability are very simple. It is however mistaken to treat epistemic probability, as some writers have done, as holding between sentences. A sentence is a grammatically well formed combination of words. A proposition is what a sentence states, what it claims about the world. In virtue of the meanings of the words in it, a sentence states the proposition which it does. But it is not the sentence S_1 'the pack has fifty-two cards, of which thirteen are hearts' which renders probable the sentence S_2 'the top card is not a heart'. If the words had different meanings S_1 would not render S_2 probable. It is what the sentence says which renders what the other sentence says probable; or what the proposition stated by the one sentence reports which renders probable what the proposition stated by the other sentence reports. In Chapter II we shall need to distinguish epistemic probability from probability of other kinds. But for the remainder of this chapter and throughout the book, unless stated otherwise, the probability referred to will be epistemic probability.

We can have propositions of classificatory, comparative, or quantitative (epistemic) probability. A proposition of classificatory probability has the form 'evidence e renders probable hypothesis h', or 'e renders h likely to be true'. ('Likely' seems to be a synonym for 'probable' and I so understand it throughout this book.) A proposition of comparative probability has the form 'evidence e renders probable hypothesis h more than evidence e^1 renders probable hypothesis h^1'. A proposition of quantitative probability has the form 'the probability of h given evidence e is p', where p is some number between 0 and 1. What is the relation between propostions of these three types? 'e renders probable h' usually means 'e renders h

more probable than e renders $\sim h$'.[1] Henceforward I will understand it in that sense. It is tempting to suppose that 'e renders h more probable than e^1 renders h^1' means 'the probability of h given e is some value p and of h^1 given e^1 is some value p^1 such that $p > p^1$' where p and p^1 are numbers between 0 and 1. However, some writers have denied that all probabilities have exact numerical values. I shall discuss the question of whether they do in Chapter III. Yet, clearly, if the probability of h given e is p, and of h^1 given e^1 is p^1 and $p > p^1$ (when p and p^1 are numbers between 0 and 1), then e renders h more probable than e^1 renders h^1. To say that the probability of h given e is 1 is to say that, given e, h is certain. To say that the probability of h given e is 0 is to say that, given e, $\sim h$ is certain. I shall represent the probability of h given e by $P(h/e)$. In this notation the classificatory propostion 'e renders h probable' becomes '$P(h/e) > P(\sim h/e)$'; the comparative proposition, 'e renders h more probable than e^1 renders h^1' becomes '$P(h/e) > P(h^1/e^1)$'; and the quantitative proposition 'the probability of h given e is p' becomes '$P(h/e) = p$'. I do not assume in the former cases that $P(h/e)$ etc. can always be given exact numerical values.

Many writers use the term 'confirms' or some similar technical term for describing the extent to which evidence renders hypotheses probable, (and hence the name, Confirmation Theory, for this subject). If we do this, it is important to keep clear the relation between the technical concept of confirmation and the more ordinary concept of probability. Failure to make clear the exact sense in which 'confirms' is being used has led to paradoxes. Following many writers I shall use 'confirms' to measure the extent to which evidence e raises the probability of hypothesis h, a use of 'confirms' which seems roughly the use which it has in ordinary language. A proposition of classificatory confirmation, 'e confirms h', will be used to mean 'the addition of e to background evidence k increases the probability of h' (where our background evidence is our evidence before we learn about e); in symbols '$P(h/e . k) > P(h/k)$'. What 'e confirms h' means will thus be unclear unless we know what the back-

(1) It is necessary to use here and hereafter symbols used in elementary deductive logic. The student unfamiliar with these is recommended to consult an elementary logic textbook, such as Patrick Suppes *Introduction to Logic*, Princeton N.J., 1957. However, as I use each new symbol I shall indicate very briefly its meaning, although in a necessarily oversimplified way, for the benefit of those who are unsure about the meaning of a particular symbol and are not in a position readily to refer to a textbook. I use in the next two pages '.' and '$\sim$'. '$\sim$' is the negation sign. '$\sim p$' says 'it is not the case that p'. The symbol '.' is the conjunction sign. '$p . q$' says 'both p and q'. '$p . q$' is true if and only if both p is true and q is true; otherwise it is false. Sometimes, instead of '.', '&' is used.

ground evidence is. Sometimes k is understood as the tautological evidence, that is, no evidence at all. In that case 'e confirms h' translates 'e renders h more probable than does no evidence at all'. 'e disconfirms h' will be used to mean '$P(h/e . k) < P(h/k)$'. 'e is irrelevant to h' will be used to mean 'e neither confirms nor disconfirms h'. A proposition of comparative confirmation, 'e confirms h more than e^1 confirms h^1', will be used to mean 'the proportion by which the addition of e to background evidence k increases the probability of h is greater than the proportion by which the addition of e^1 to background evidence k increases the probability of h^1. (These proportions may be $>$, $=$, or < 1); in symbols $\dfrac{P(h/e . k)}{P(h/k)} > \dfrac{P(h^1/e^1 . k)}{P(h^1/k)}$. A proposition of quantitative confirmation, 'the confirmation of h given e is c' will be used to mean 'the ratio of the probability of k given e and k to its probability given k alone is c': in symbols,

$$\frac{P(h/e . k)}{P(h/k)} = c.$$

The two triplets of classificatory, comparative and quantitative uses of concepts of 'probability' and 'confirmation' which I have described correspond to the two triplets listed by Rudolf Carnap in the Preface to the second edition of his *Logical Foundations of Probability* ([1] p. xvi), a classic work on the subject. He called these two triplets 'three concepts of firmness' and 'three concepts of increase in firmness'. However, as he admits in this Preface, the three concepts which he uses in the main part of the book and there calls the concepts of classificatory, comparative and quantitative confirmation are what I have called the concepts of classificatory confirmation, and comparative and quantitative probability respectively. (He sets these out in Chapter II of [1].) These three do not fit together to make a neat set.

So this book seeks to expound the problems of Confirmation Theory, to state and criticize the views of philosophers about these, and in the process to contribute towards the construction of a correct Confirmation Theory. I understand by a correct Confirmation Theory one which draws out the criteria implicit in our ordinary judgements of epistemic probability, which we all recognize as correct.

There are many past cases in the history of science or other history (in a very wide sense) where we all unhesitatingly judge that on the evidence available one theory or one prediction was more likely to be true than another. Further, there would seem to be an infinite number of imaginary cases which we can construct where some theory of prediction is, intuitively, more probable or more likely to be true than another or more probable on new evidence than on previous evidence, or has on certain evidence such and such a degree of probability. These cases form our data from which we seek to draw out criteria of confirmation.

Probability on evidence, or epistemic probability, may be ascribed to propositions of all kinds – moral, aesthetic, and theological propositions, as well as historical and scientific ones. However, as is traditional in books on probability, I will confine myself to historical and scientific propositions (in wide senses of 'historical' and 'scientific'). My justification for doing this is that men are in general agreed about many cases of these kinds, where and to what degree evidence renders hypotheses probable. All men who have studied the issues agree that the deflection of light by the sun in 1919 confirmed the laws of the General Theory of Relativity; that the discovery of a man's fingerprints confirms (for normal background evidence) the proposition that he was present there; that if all hundred throws of a coin have resulted in throws of heads, then (in the absence of other relevant evidence) it is considerably more probable that the next throw will result in heads than in tails. Here we have clear cases of one hypothesis being confirmed, or made more probable than another one, from which we can draw out criteria of confirmation. But in morals or theology there are too few examples of agreed judgements of confirmation to provide data for a Confirmation Theory. Yet once a Confirmation Theory has been established for scientific and historical propositions, we can then try to apply it to propositions of other kinds to ascertain which of those are rendered probable by which evidence, given those criteria of confirmation which are used in science and history.

Nevertheless, although in general men are agreed about when and to what extent a historical or scientific proposition is rendered probable by evidence, there are cases where men do not agree, and, worse, several whole kinds of cases about which men disagree. We shall meet various such kinds of cases. In general, the aim of this book is to record the vagueness and impression involved in our ordinary criteria of epistemic probability, rather than to lay down standards more precise than those in common use. However, sometimes we shall find it valuable to draw out the consequences of one of two conflicting views found in common use. I shall make it quite explicit when I am doing the latter and give my grounds for doing so.

In setting out for consideration various criteria of confirmation which have been proposed, it will be necessary, as we have seen, to justify them by appealing either to actual cases from history or science or to imaginary cases, where we agree about the judgements of confirmation which we would make. However, in order to keep this book to a reasonable size, there will not be a great number of detailed examples from the actual history of science. To describe these examples in sufficient detail to draw claims about probability from them takes up much space. There will, in contrast, be a number of imaginary examples, where we postulate that we have only a small amount of evidence which is relevant for assessing the probability of a certain hypothesis, and that we have no other evidence which makes any difference to the probability of the hypothesis under consideration, and so all other evidence can be ignored. From time to time we shall remind the reader of the existence of this other evidence. These examples are only extreme forms of normal cases where the amount of evidence we have which *crucially* affects the probability of some hypothesis is fairly small compared with the vast amount of other knowledge which we have about the world.

I have written that we seek to draw out from such cases, actual and imaginary, *our* criteria of confirmation. But it can be claimed, and has been argued persuasively by Kuhn [6], that criteria of confirmation vary from age to age. While one generation might consider that a certain e confirmed a certain h, for background evidence k, another generation might reach a different judgement. This could be, though it may be that Kuhn has exaggerated the transitoriness of criteria of confirmation. However, I am only seeking to draw out the criteria of confirmation of this generation and culture. I shall use historical and imaginary examples, and ask (e.g.) which of rival hypotheses h or h^1 was best confirmed by e for background evidence k, where k is a very different set of assumptions about the world from those which we make. But I am asking what we, with our criteria of rationality, say about h and h^1, *given e and k*. Sometimes the answer will be clear, sometimes not.

My concern is with fundamental principles of confirmation, such as whether all propositions about probability conform to the axioms of the probability calculus. Some confirmation theorists have been concerned not merely to lay down such fundamental principles, but to develop a formal theory of confirmation in which theorems expressed in symbols are derived from such principles by many steps of rigorous deductive inference. The most famous formal Confirmation Theory is that of Rudolf Carnap. His first and most quoted work is *The Logical Foundations of Probability* [1]. It was followed up by the small *The Continuum of Inductive Methods* [2], and by the later work edited with R.

Jeffrey, *Studies in Inductive Logic and Probability. Volume I* [3].
Another formal Confirmation Theory is the system outlined in a
number of articles written since 1965 by Jaako Hintikka (see for
example [5]). I shall not be concerned in this book with the details of
such theories, but will discuss the fundamental principles which they
use.

In considering the epistemic probability of propositions of science or
history I shall confine myself to three most primitive and interesting types
of proposition functioning as evidence or hypothesis – particular propositions, general propositions and nomological propositions – and combinations of these. (It is arguable that all synthetic propositions of other
kinds in science or history – e.g. causal propositions – can be analysed in
terms of propositions of these three types.)

Propositions of the three cited types, as I shall define them, are all
synthetic or empirical propositions. A proposition p is synthetic if the
denial of p (the proposition '$\sim p$') makes sense, describes a possible state
of affairs. Thus 'my desk is black', 'Mr Churchill became Prime Minister
in 1940', and 'the moon is made of green cheese' are all synthetic propositions, the first two true, the third false. In contrast, a proposition p is
analytic, or logically true, or true of logical necessity, if the denial of p
is self-contradictory, that is does not describe a logically possible state
of affairs, a state of affairs the description of which makes sense. Thus
'all bachelors are unmarried', 'a gorilla is an animal', '$3 + 5 = 8$' are analytic propositions. Particular propositions are synthetic propositions
which predicate something of, that is ascribe a property to, one particular object at some time or times. All the three synthetic propositions
cited above are particular propositions.

General propositions are synthetic propositions which ascribe some
property to all or some stated proportion of members of some class at
some time or times. Thus '90% university teachers vote Labour' or 'all
the inhabitants of Fisher Place are Irish' are general propositions. If the
general proposition has the form 'all A's are B' or '100% A's are B', it
will be a universal proposition; if it has the form '$n\% A$'s are B', where
$n \neq 100$ or 0, it will be a statistical proposition. ('0% A's are B' means
the same as 'all A's are not – B'.) General propositions just say that some
property in fact holds of certain objects. Their general form being '$n\%$
A's are B', they will be true if there are A's and $n\%$ of them are B's; they
will be false if there are A's and it is not the case that $n\%$ of them are B's;
they will be neither true nor false, if they fail of reference, that is if there
are no A's.

Nomological propositions are propositions which claim that of physical necessity a certain proportion of objects of a certain kind '*A*' have some property '*B*'. The most usual type is the universal type, that is where the proposition says that of physical necessity all *A*'s are *B* or that of physical necessity all *A*'s are not - *B*; but we shall meet in the next chapter statistical nomological propositions also. Like general propositions, nomological propositions may be written as (e.g.) 'all *A*'s are *B*', or even just '*A*'s are *B*', but in their case this is elliptical for 'of physical necessity all *A*'s are *B*'. Natural laws have this form. 'Light travels at a constant velocity, *c*, *in vacuo*' (*c* = 300,000 km/sec) put forward as a law of nature says 'of physical necessity all light travels at a constant velocity, *c*, *in vacuo*'. 'Silver dissolves in nitric acid' when put forward as a law of nature is elliptical for 'of physical necessity all silver dissolves in nitric acid'. The difference between nomological and general propositions can be seen in two ways. First nomological propositions entail counterfactuals while general propositions do not. (A counterfactual is a conditional proposition in which it is presupposed that the antecedent is false. 'If Hitler had won the war, we would all now be slaves' is such a proposition for it presupposes that Hitler did not win the war. 'If Jones went by train, he will now be home' is not a counterfactual, for it does not presuppose that Jones did not go by train.) The general proposition 'all *A*'s are *B*' does not entail counterfactuals of the form 'if *a* had been *A*, it would have been *B*' (when *a* is some individual object) whereas 'of physical necessity all *A*'s are *B*' does entail 'if *a* had been *A*, it would have been *B*'. 'All men living in Alban Street have black hair' does not entail 'if I had been living in Alban Street, I should have had black hair'. But 'of physical necessity silver dissolves in nitric acid' does entail 'if this silver had been put into nitric acid yesterday it would have dissolved'.

Secondly, as we have noted, general propositions of the form '*n*% *A*'s are *B*' will only be true or false if they have reference, that is if there are any *A*'s. If there are no inhabitants of Fisher Place, it is neither true nor false that all the inhabitants of Fisher Place are Irish. However, nomological propositions, that is propositions of the form 'of physical necessity *n*% *A*'s are *B*', may be true or false even if there are no *A*'s. Newton's first law can be written as 'of physical necessity all bodies under the influence of no force remain at rest, if they are at rest; and continue with uniform velocity, if they are in motion'. There are no bodies under the influence of no force, for all bodies are subject to the force of gravity, yet the law is true. Scientists frequently dispute about the properties of non-existent objects which never have existed and never will exist - bodies falling in a perfect vacuum, ideal gases, elements with enormously high atomic numbers etc. False and true nomological propositions can be made about such objects. For instance it would be false to claim that all ideal gases have $pv = RT^3$ (when p is the pressure, v the volume, T the temperature of a gas, and R is the gas constant).

So although nomological propositions may, like some general propositions, be written '$n\% A$'s are B', and most usually 'all A's are B', they differ in meaning from them. Scientific theories are bodies of nomological propositions. They tell us not merely what does happen but what would happen under various circumstances.

In this chapter I have outlined the subject matter of Confirmation Theory and explained the meaning of 'epistemic probability', 'confirmation', and other technical terms which will be of subsequent use. In the next chapter I shall need to distinguish epistemic probability from probability of other kinds and in this context to consider various theories about the meaning of 'probability'. In Chapter III I shall expound the calculus of probability, and in that chapter and the next three chapters I shall examine and defend the claim that it correctly states logical relations which hold between propositions of epistemic probability. In Chapters VII to XI I shall use these results to consider problems of how to assess the probability of propositions of certain kinds (e.g. nomological propositions) on certain kinds of evidence. In Chapters XII and XIII I shall contrast the probability of propositions with their acceptability, and discuss criteria for the acceptability of propositions.

Bibliography

The most famous builder of a Confirmation Theory is Rudolf Carnap. For his theory, see:

[1] RUDOLF CARNAP *The Logical Foundations of Probability* (first published 1950), 2nd edition, London, 1962.

[2] RUDOLF CARNAP *The Continuum of Inductive Methods,* Chicago, 1952.

[3] R. CARNAP and R. C. JEFFREY (eds.) *Studies in Inductive Logic and Probability. Volume I,* London, 1971.

For Carnap's view on the method to be adopted in constructing a confirmation theory, see Prefaces and Chapter I of [1] and §18 of [2]. For criticism of these views and fuller exposition of the points made in this chapter about how to construct a Confirmation Theory, see:

[4]　R. G. SWINBURNE 'Choosing Between Confirmation Theories' *Philosophy of Science*, 1970, **37**, 602–13.

For Hintikka's theory, see:

[5]　JAAKO HINTIKKA 'A Two-Dimensional Continuum of Inductive Methods' in Jaako Hintikka and Patrick Suppes (eds.) *Aspects of Inductive Logic*, Amsterdam, 1966.

For Kuhn's claim about the transitoriness of standards of confirmation see:

[6]　THOMAS S. KUHN *The Structure of Scientific Revolutions*, Chicago, 1962.

For the distinction between general and nomological propositions (which he calls the distinction between accidental generalizations and universal laws), see:

[7]　C. G. HEMPEL *Philosophy of Natural Science*, Englewood Cliffs, N.J., 1966, pp. 54–8.

CHAPTER II

Physical and Statistical Probabilities

The terms 'probable' and 'probability' have other uses than to express the extent to which one event provides evidence for the occurrence of another, or one proposition provides evidence for the truth of another. When 'probable' and related words are concerned with the latter, I have said that they express propositions of epistemic probability. Epistemic probability is the main concern of this book, and it is important at this stage to distinguish it from two other kinds of probability, physical probability and statistical probability. Physical probability is a measure of the extent to which an event is determined by causes. In this sense, to say that the probability of an event S is 1, is to say that, given the factors which affect whether S occurs or not, S could not but occur. Given that causes precede their effects, this is to say that, given the precedent states of the universe, S could not but occur. Conversely, to say that the probability of an event S is 0 is to say that, given the precedent states of the universe, S could not occur. The nearer to 1 is the probability of S, the greater is the tendency of the preceding states of the universe to bring about S. The physical determinist claims that the probability, in the sense of physical probability, of all events is 1 or 0. The indeterminist claims that some events have physical probabilities intermediate between 1 and 0. A man who believes that the statistical laws of Quantum Theory are basic laws of nature is an indeterminist. For he believes that the probabilities which Quantum Theory tells us about, such as that an atom of carbon-14 has a probability of 1/2 of disintegrating within 5,600 years, are intrinsic to material objects, being not further explicable. Hence he believes that the physical probability of a particular atom of carbon-14 disintegrating within 5,600 years is 1/2, and so is other than 1 or 0.

If determinism is true, the physical probability of all events will be 1 or 0, but the epistemic probability of an event and so of the proposition reporting the event may still take any value between 1 and 0, dependent on the evidence available. Conversely, if indeterminism is true and the physical probability of a particular atom disintegrating within a certain 5,600 years is 1/2, the epistemic probability, on evidence available after the period, of its having disintegrated within the period could be 1 or 0, depending on the evidence.

Epistemic and physical probability must both be distinguished from statistical probability. Propositions of statistical probability are propositions about proportions or relative frequencies. The typical form of such a proposition is 'the probability of *an A* being *B* is *p*'. (Note the indefinite article. No one particular *A* is referred to.) Examples of such propositions are 'the probability of throwing a seven with two dice is 1/6' and 'the probability of an inhabitant of New York voting Republican is 0·53'. That these are not propositions of epistemic probability can be seen by the fact that when the relevant evidence available to us is set out and rationally assessed, we could still be mistaken in our probability judgements. Having studied a sample from the population, I may reasonably claim that the probability of an inhabitant of New York voting Republican is 0·53, but admit that I could be mistaken, that in fact the probability could be much higher. Subsequent observation will reveal whether or not I am mistaken. But it is characteristic of epistemic probability that (once any implicit reference to evidence has been made explicit, and the evidence fully stated) what is the probability conferred by certain evidence on a certain hypothesis is not an empirical matter, a matter of how things in the world in fact behave, which could be settled by observation of any kind. If I assess the evidence available to me now properly and conclude that on that evidence the probability of Jones having voted Republican at the last election is such and such, subsequent discoveries (including discovery for certain whether or not Jones voted Republican) cannot affect the value of that probability, cannot show that I was mistaken in my judgement of it. Or if I conclude today from study of the evidence now available that Quantum Theory is probably true, subsequent empirical discoveries may show Quantum Theory to be false, but they cannot show that it was not probable on today's evidence. Therefore such propositions as 'the probability of an inhabitant of New York voting Republican is 0·53' are to be distinguished from propositions of epistemic probability. The former proposition rather says (roughly) that 53% New Yorkers vote Republican. 'The probability of throwing a seven with two dice is 1/6' says (roughly) that when two dice are thrown sevens turn up a sixth of the time. As statistical probabilities are of crucial importance for calculating epistemic probabilities, a more detailed analysis of propositions of statistical probability will be provided shortly.

Although propositions of physical and statistical probability are not *about* the relation of evidence to hypothesis, they can of course be and normally are asserted on the basis of evidence. We can say on the basis of our evidence *e* that the probability of an *A* being *B* is 0·05 or that the physical probability of *S* occurring is 0·007. But there is all the difference in the world between asserting on evidence *e* that *S* is probable and asserting that on evidence *e S* is probable. When we assert a proposition of epistemic probability, we are doing the latter, asserting a proposition *about* the relation of evidence to hypothesis.

Clearly it is no accident that the same word 'probability' has been used to describe epistemic, physical, and statistical probability. For one thing, as we shall see, the axioms of the probability calculus apply to propositions of physical and statistical probability, and at any rate to many propositions of epistemic probability (and, I shall claim, to all such propositions). For another thing the values of the different probabilities relevant to particular situations are often the same. Thus, if the evidence *e* about *S* consists of a full description of the world before the time referred to in *S*, the epistemic probability of the occurrence of *S* given *e* will have the same value as the physical probability of the occurrence of *S*. But even if the two 'probabilities' have the same numerical value, the claim that the 'probability' has that value has a different meaning according to the kind of probability in question. To say that certain evidence makes a certain proposition likely to be true to such and such an extent is not the same as to say that certain causes have such and such an influence on whether some event happens. Again, if we are ascertaining the probability of an object *a* having a property *Q*, and our evidence is that *a* has the property *A* and that the proportion of *A*'s which are *Q* is *p*, then, intuitively, the epistemic probability that *a* is *Q* on that evidence will have the same value as the statistical probability of an *A* being *Q*, that is *p*. If our evidence is that 10% *A*'s are *Q* (i.e. the statistical probability of an *A* being *Q* is 0·1) and that *a* is *A*, then the epistemic probability on that evidence of *a* being *Q* is 0·1. But, as before, if two functions have sometimes the same value, it does not follow that they are the same function. The epistemic probability that *a* is *Q* will have values where no relevant statistical probabilities are known. Let *a* be a certain throw of a certain coin *C*. Let the sole evidence be that *a* is *A*, a throw of *C* which is a coin biased by having either two heads or two tails. The epistemic probability of *a* being a throw of heads, *Q*, on the stated evidence is surely 1/2. We can say this without having any information about the statistical probability of an *A* being *Q* or about any other statistical probability. We can even say it if we *know* that the statistical probability of an *A* being *Q* is other than 1/2, e.g. as in this case, that it is either 1 or 0. In one sense of probability, the statistical, the probability of a throw of *C* being heads is either 1 or 0. In another sense of probability, the epistemic, the probability of a certain throw of *C* being heads is 1/2; for what is meant here is that the probability that *a* will be heads, where *a* is a certain throw of coin *C*, on evidence that *C* is biased in the way stated, is 1/2. What the previous sentence said was that the frequency of heads among throws of the biased coin is either 1 or 0.

Over the past two hundred years philosophers have put forward five well known theories of probability, each theory having many variants. They are, in historical order, the Classical, Frequency, Logical, Subjective and Propensity Theories. These theories seem usually to be intended as analyses of the concept of probability, that is, accounts of what is meant when something is said to have such and such a probability. We must see how far these theories give an adequate account of the three kinds of probability which we have distinguished.

I will deal with the Frequency Theory first. This theory was first put forward by Venn [2], and was developed more fully by Von Mises [3]. More sophisticated versions are due to Reichenbach [4], Braithwaite [5] and Salmon [6]. This theory claims that propositions about probability are propositions about proportions or frequencies. Clearly, as we have seen, some probability propositions are of this kind, and these I have called propositions of statistical probability. But for the reasons given above propostions of epistemic probability are not of this kind. A detailed Frequency Theory can best be regarded as a detailed analysis of probability propositions of one kind, viz. propositions of statistical probability. As statistical probability is of great importance for the calculation of epistemic probability, we must provide a more detailed analysis.

Propositions of the form 'the probability of an *A* being *B* is *p*' seem to be of two kinds. A proposition of one kind is analysable as a proposition about actual proportions; a proposition of the other kind is analysable as a proposition about hypothetical proportions. I will call propositions of the first kind simple statistical probability propositions and those of the latter kind complex statistical probability propositions. I do not think that grammatical form always provides a clear guide to logical form, of which kind a statistical probability proposition is, but some rough indications can be given. Often, I suspect, someone who utters a proposition of the form 'the probability of an *A* being *B* is *p*' is unclear about exactly which kind of proposition he is uttering.

If 'the probability of an *A* being *B* is *p*' is a simple statistical probability proposition, it is simply saying that the proportion of *A*'s which are *B* is in fact *p*. This analysis seems plausible for propositions where the '*A*' is such that of logical necessity there is one definite number *n* of *A*'s. An example of this is where '*A*' is 'a card in this normal pack of cards'. To say that the probability of a card in this normal pack of cards being a heart is 1/4 is just to say that 1/4 of the cards in this normal pack are hearts. This analysis also seems plausible for cases where there is a specific time reference contained in '*A*', so that of logical necessity an object can only be an *A* at a certain temporal instant or during a certain temporal interval. This will be so if '*A*' is 'an inhabitant of Fisher Place on 1 January 1970' or 'a man born during 1950'. To say that the probability of an inhabitant of Fisher Place on 1 January 1970 being under 14 years of age is (or was) 0·2 seems to be to simply say that 20% inhabitants of Fisher Place on 1 January were under 14 years of age.

If there is no specific time reference in '*A*' there does normally seem to be a time reference involved in the probability proposition as a whole. 'The probability of a university teacher owning a TV set is 0·9' says that the probability is *now* 0·9; it allows that the probability in question might have been 0·5 ten years ago. However where the time reference is contained in the verb of the probability proposition as opposed to being contained in the antecedent predicate (the '*A*'), and especially if it is made by 'is' as opposed to 'was' or 'will be' followed by a specific date, the above analysis will not always work. Certainly it seems plausible for some cases. To say that the probability of a university teacher owning a TV set is 0·9 seems to be to say that 90% university teachers now own TV sets. But there are other cases where it clearly is not satisfactory. In particular it does not seem to work in cases where '*A*'s' are things which occur in series – e.g. 'tosses of this coin' – and the actual number of *A*'s is small. In such a case if the temporal reference is given by the 'is' of the probability proposition – 'the probability of a toss of this coin being heads *is* 1/2' – the reference must be to a vague period of time which includes the present instant (for there can be at most one toss of this coin occurring at exactly this present instant). But now suppose I have a new coin, and it has just been tossed twice, both occasions resulting in a fall of heads. On the above simple analysis, the probability of a throw of heads is the number of occasions on which the coin has landed or will land heads during some period including the present instant divided by the total number of throws of the coin. In this case if the coin is destroyed before it is tossed again, the probability of a toss of the coin being heads is 1, whereas if we toss it eight more times before destroying it, and six of the total number of throws results in heads, four in tails, then the probability of a throw of heads is 6/10. It seems absurd, that is to run contrary to our understanding of what is meant by probability, to make the probability depend on whether we destroy the coin. Again suppose the coin was destroyed before being tossed at all; on the simple frequency analysis there would have been no probability of a throw of the coin being heads, which seems very odd.

It seems natural to meet this difficulty by a long-run theory and to suppose that in such a case where A's occur in succession and the number of A's is small, that the probability of an A being B is the proportion of A's which would be B if there were many more A's, that is the frequency of B's among A's in the long-run. But this is very vague. How 'many more A's'? How long is the 'long-run'? To make this theory more precise it is tempting to turn it into an infinitist theory. Such a theory was put forward by Von Mises [3] and developed by Reichenbach [4]. On this view the probability of an A being B, where n is the number of A's and r the number of A's which are B, is given by $\lim_{n \to \infty} (r/n)$. To say that $\lim_{n \to \infty} (r/n) = p$ is to say that for every finite real number $\delta > 0$, however small, there is some number of A's n_δ such that for any $n \geqslant n_\delta$, $p + \delta \geqslant r/n \geqslant p - \delta$. In less formal terms this is to say that if we were to take enough A's we would find that the proportion of A's which were B remained as close as we liked to p. (Von Mises's [3], Chapters I and III, added the further qualification that the series be random.)

So, on this view, to say that the probability of a throw of this coin being heads is 1/2 is to make a claim about the result of an infinite series. But such a claim is compatible with anything past or future which could be observed. Even if each of a million throws of the coin has resulted in a throw of heads, it is on this view still compatible with observations to say that the probability of a throw of heads is 0. Adopting a claim about the result of an infinite series does not commit a man to the truth or falsity of any predictions which could be tested. So a claim about the result of an infinite series seems rather empty. Let us therefore return to the somewhat vague concept of the long-run and try to see what is involved in it. On this view to say that the probability of an A being B is p is to say that the proportion of B's among A's would be p, if there were many more A's. The probability statement thus involves a hypothetical element.

Passing over for the moment the difficulty about the vagueness of this claim, what counts as 'many more', let us deal with another difficulty (a difficulty which also arises on the infinite series account). To say that the probability of a throw of a certain coin being heads is 1/2 is, on this view, to say that if there were many more tosses, the proportion of tosses which were heads would be 1/2. But if the series were continued, some continuations of it would result in all heads, e.g. if the table on which the coin falls and the tail side of the coin were made into opposite poles of magnets. So to meet this difficulty one has to add to the definition 'if the series were continued under similar conditions'. But what constitutes 'similar conditions'? Presumably (and certainly if determinism is true), if heads has been thrown first time, and the conditions are *sufficiently* similar on all subsequent occasions (air pressure, tilt of the coin when thrown, etc.) heads would occur every time. Yet we would not for this reason wish to say that the probability of a throw of heads was 1. Clearly some variations in initial conditions are allowed. But which variations? What seems to be involved is that the variations should occur in the proportions in which they occur for events of that kind in the spatio-temporal region for which we are assessing probability. In this case, where we are investigating the probability of a toss of a specified coin having a certain property, we are investigating how many tosses of the coin would have had that property under a variety of initial conditions, proportional to the actual frequency of such conditions for similar events, i.e. tosses of other coins. If coins are tossed more often beginning from a horizontal position than from a position slanting to the vertical, this is supposed to be so in the distribution of hypothetical initial conditions for tosses of our coin. If coins are normally tossed so as to land on to flat surfaces, this too is supposed to be so in the distribution of hypothetical initial conditions for tosses of our coin. But, in a deterministic universe, the number of tosses of a coin which would result in heads under a variety of initial conditions under which similar events occur, proportional to their actual occurrence, will be independent of the actual number of tosses. If the proportion of initial conditions for which tails would result to the proportion for which heads would result is kept constant, the proportion of tosses of a coin which would be tosses of heads is independent of the actual number of tosses.

These considerations suggest the following second analysis of statistical probability. This kind of statistical probability I will call complex statistical probability. In this analysis I take account of the fact that the universe may not be deterministic, that is, the physical probabilities of all or some events may be other than 1 or 0.

We are assessing the probability of an A being B. Let us denote by $A*$ the causal ancestor of an A; that is, an object is $A*$ if it is not an A but such as develops into an A. $A*$'s could occur under a variety of initial conditions. Let us call all initial conditions I_1-conditions for which of physical necessity any $A*$ which occurs under one of those conditions becomes an A which is B. Let us call all initial conditions I_0-conditions for which of physical necessity any $A*$ which occurs under one of those conditions becomes an A which is not-B. Let us group all other initial conditions into classes for which the physical probability of an $A*$ becoming an A which is B is the same for all members of the class but different from what it is for members of other classes. Thus an initial condition will be an $I_\frac{1}{2}$-condition if the physical probability of an $A*$ becoming an A which is B under that initial condition is $\frac{1}{2}$. Now I shall take it as axiomatic that if the physical probability of an $A*$ becoming an A which is B is p_1, then in a class of n $A*$'s, that number which it is physically more probable will become A's which are B than that any other particular number will is denoted by the nearest integer to np_1. (If np_1 lies midway between two integers q and $(q + 1)$, then it is physically equally probable that there will be q A's which are B as that there will be $(q + 1)$ A's which are B, and physically more probable that there will be q A's than that there will be any other particular number apart from $(q + 1)$.) At the end of this chapter I shall prove a similar proposition for statistical probability, but as I am conducting no detailed analysis of physical probability, I ask the reader to take this as intuitively obvious. In this case, for large n, the most probable proportion of B's among A's, to a degree of approximation which we can ignore, will be p_1. Now let us construct an imaginary collective, which I shall call a P-population, as follows. Initial conditions under which $A*$'s could occur, occur in the proportions in which they do for causal ancestors of objects of the type of A in the relevant spatiotemporal region of the actual world. If I_1's occur twice as often as I_0's in the actual world, then we imagine a collective in which $A*$'s occur under the various possible initial conditions, I_1's occurring twice as often as I_0's. Many $A*$'s occur under each type of initial condition. For each type of initial condition, the most probable proportion of $A*$'s to become A's which are B does so. Now, I suggest, to say that the complex statistical probability of an A being B is p is to say that the proportions of $A*$'s which would become A's which are B in the hypothetical P-population is p. There seems no reason why the physical probability of any event cannot take as a value any of the real numbers p, $0 \leqslant p \leqslant 1$, and hence a complex statistical probability could also have any such value.

This analysis presupposes that it is clear what are 'objects of the type of A', that is of what genus A is a species, that tosses of this coin are a species of tosses of coins, university teachers are a species of employees, etc. Sometimes this will not be clear and here the probability proposition will have a corresponding inexactness. It also presupposes, if the temporal reference is vague, e.g. to a period including the present instant with boundaries not specified, that shifting the boundaries somewhat does not produce great change in the distribution of initial conditions; for if it did the vagueness about the boundaries of the temporal period produces corresponding vagueness in the probability. This analysis of statistical probability may sometimes apply even when the actual number of A's is large, although usually in such a case, as I pointed out earlier, analysis in terms of actual proportions seems correct. Why in some cases where the number of A's is large we suppose that the probability of an A being B is simply the number of A's which are B is because if the number of A's is large we reasonably suppose that their causal ancestors occurred under a variety of initial conditions typical of those to be expected for objects of its type. However, where we have reason to believe that this is not so we sometimes withdraw the claim that the probability is the same as the actual frequency. Suppose that before being destroyed a certain coin has been tossed many, many times by a careful experimenter who has always placed it tail upwards and exactly the same distance above the table on which it is to land before tossing it. Suppose that 70% of the large number of tosses of the coin have been tails. Is the probability of a toss of that coin being tails in that case 0·7? We might be inclined to say not, on the grounds that the frequency results were obtained under untypical conditions. We might be prepared to admit that actual frequency only shows probability if the coin had been tossed under a typical variety of initial conditions. If that was our reaction, we would be treating 'the probability of a toss of this coin being heads is 0·7' as a complex statistical probability proposition.

So I would suggest that there are statistical probability propositions of two kinds – simple and complex. Simple statistical probability propositions concern actual frequencies; they state the actual proportion of A's which are B, and so are general propositions. In this connexion I shall find it useful to use later the following notation. Where a number

n of A's have been observed and it has been observed how many of them are Q and how many not-Q, I shall denote by '$\Phi(Q/A)$' the proportion of A's found to be Q. $\Phi(Q/A)$ will thus be the simple statistical probability of an observed A being Q. I shall sometimes add a subscript below the Φ to show the number of A's which have been observed. '$\Phi_n(Q/A) = p$' will be used to mean 'the proportion of Q's among the n observed A's is p'. Complex statistical probability propositions on the other hand concern hypothetical frequencies, frequencies in an imaginary P-population. Complex statistical probability propositions are thus statistical nomological propositions, although ones of a highly complex character. They have the form '$n\% A$'s in a P-population are of physical necessity B'. Given the character of the population it is a matter of physical necessity what proportion of A's in it will be B. I shall denote in future by $Pr(B/A)$ the complex statistical probability of an A being B. If $Pr(B/A)$ equals p whatever the proportions in which initial conditions occurred in the actual world, then p is the physical probability of an A being B. Thus if of physical necessity all A's are B and so the physical probability of an A being B is 1, then whatever the initial conditions under which A's occur, $Pr(B/A) = 1$.

It will not necessarily be clear, I have claimed, from the grammatical form of a statistical probability proposition of which kind it is. However, I have given some rough suggestions for determining to which kind some statistical probability propositions belong. Sometimes indeed the utterer may not always be clear about the meaning which he is intending to give a sentence of the form 'the probability of an A being B is p'. Sometimes he may mean something vague somewhere in meaning between a simple and a complex statistical probability proposition. However, quite often, I suggest, people intend to give to a sentence 'the probability of an A being B is p' one of the two meanings which I have described. They seem also the senses of statistical probability most useful for Confirmation Theory.

Neither of the analyses provided is adequate for the case where the number of A's is actually infinite. It may be, though scientific evidence is not very definite here, that the number of electrons or the number of stars is infinite; which is to say that however many you count, you can still find another. If the number of A's is infinite, the probability of an A being B cannot be given by the proportion of A's which are B or would be B under certain circumstances, for there is in that case no proportion. However, as my primary concern is not with statistical probability, and it is open to question whether there are any A's (where 'A' denotes a material object of a distinct kind) of which the number is infinite, I shall not provide a theory of these. (Apart from limiting frequency analyses, such as those of Von Mises [3] and Reichenbach [4] referred to earlier, there is the analysis of such propositions by Kneale [13] in his range theory.) As far as my subsequent analysis is concerned, no harm will result from treating propositions about the probability of electrons having some property as propositions about the probability of electrons within the observable universe (of which there will be a finite number) having that property.

The Frequency Theory of probability is an empirical theory in that according to it probability propositions are empirical or synthetic propositions (that is, propositions making claims about how things are in the world, which it makes sense to suppose to be true or to be false of it). They are shown to be reasonable or unreasonable to assert by the actual behaviour of physical objects. A different empirical theory of probability is Propensity Theory. A version of it is due to Karl Popper ([7] and [8]) and other versions of it have been put forward by Hacking [9] and Mellor [10] (though Hacking could be classified as a Frequency theorist). Propensity Theory claims that probability propositions make claims about a propensity or 'would-be' or tendency in things. In the simplest version of Propensity Theory, to say that a certain atom has a probability of 0·9 of disintegrating within the next minute is to make a statement about a propensity in it, which is a property of it in the same way as is its mass or charge. Now this seems the correct analysis of propositions of physical probability. If the laws of Quantum Theory are fundamental laws of nature (laws which operate whatever the initial conditions), and the cited proposition is a consequence of them, then the atom has a propensity to (is inclined to, tends to) disintegrate, but has also some propensity not to disintegrate, and its propensity to disintegrate can be given a numerical value – 0·9.

Propositions of physical probability may be particular, general, or nomological propositions. Taken as a physical probability proposition, 'the probability of this atom disintegrating within 5,600 years is 1/2' is a proposition ascribing a certain property, like charge or mass, a propensity to disintegrate, to an individual atom, and so is a particular proposition. Taken as a physical probability proposition 'all the atoms in this container have a probability of 1/2 of disintegrating within 5,600 years' is a general proposition. Taken as a physical probability proposition 'all atoms of carbon-14 have a probability of 1/2 of disintegrating within 5,600 years' is a nomological proposition. It is a law of nature, derivable within a scientific theory, and so elliptical for 'of physical necessity all atoms of carbon-14 have a probability of 1/2 of disintegrating within 5,600 years'.

I shall henceforward represent the nomological physical probability proposition 'of physical necessity all A's have a physical probability p of being B' by '$\Pi(B/A) = p$'. 'Of physical necessity all A's are B' will thus be represented by '$\Pi(B/A) = 1$'. For all p, if $\Pi(B/A) = p$, then necessarily $Pr(B/A) = p$, for if a proportion p of A's are B in any P-population, whatever the proportion in which initial conditions occur in the actual world, then it will be p for a certain distribution of them, viz. that distribution in which they do in fact occur. However, unless 'A' is a very full description of an event, there will seldom be values of $\Pi(B/A)$. There is no physica probability of a toss of a coin landing heads, because whether or not it lands heads depends on the circumstances of its tossing. But there is a physical probability of a coin tossed in a certain way with certain air pressures and a certain kind of surface landing heads. '$\Pi(B/A) = p$' entails '$Pr(B/A) = p$'. But '$Pr(B/A) = p$' does not entail any proposition of the form '$\Pi(B/A) = q$'. On this account $Pr(B/A) = 1$ could be true either because of physical necessity all A's are B (that is, $\Pi(B/A) = 1$) or because for all the initial conditions in which objects of the type of $A*$'s are to be found, all A's are B.

Although satisfactory for propositions of physical probability, Propensity Theory does not give a satisfactory account of propositions of epistemic probability. Suppose that I have before me a pack of playing cards with their faces downwards. On the evidence that it is a pack of normal make-up, the probability of the top card being a heart is 1/4. Yet to say this cannot be to say that the top card has some propensity or tendency to be a heart. It either is or is not a heart and has no propensity to be anything other than what it is.

So Frequency Theory can be seen as giving a correct account of one kind of probability statement, and Propensity Theory as giving a correct account of another kind. Neither Classical Theory nor Subjective Theory seem to me to give a correct account of any sort of probability statement. As the main concern of this book is with epistemic probability, I will justify this claim only to the extent of showing that they do not give correct accounts of epistemic probability.

The calculus of probability (which I shall set out in Chapter III) was first developed in the seventeenth and eighteenth centuries as a device for calculating the probability of events in games of chance. The understanding of probability adopted by those who developed the calculus is

known as the Classical Theory of probability. The most mature statement
of the theory is contained in Laplace's *Philosophical Essay on Probabilities*
[1]. According to Classical Theory, to say that the probability of an event
S (or of a proposition reporting *S*) is *p* is to say that there are integers
m and *n* such that $p = m/n$, and such that one of *n* exclusive and exhaustive alternatives must occur, *m* of which constitute the occurrence of *S*.
Thus to say that the probability of throwing an even number when a die
whose faces are numbered 1, 2, 3, 4, 5, 6 is thrown is 0·5 is to say that
$0·5 = m/n$, where there are *n* possible throws of the die of which *m* are
throws of an even numbered face. It looks as if this statement is true
because there are six possible throws of the die, six possible faces on which
it can land, of which three are even numbered faces.

As stated so far, the theory is not specific enough, for according to
how the exclusive and exhaustive alternatives are selected, any probability is possible for any event. We could in the above die-throwing
example describe the alternatives as each of the die faces, in which case
$\frac{3}{6} = 0·5$ would constitute the probability of a throw of an even number.
Or we could describe the alternatives as throwing 1, 3, 5 or an even number, in which case $\frac{1}{4} = 0·25$ would constitute the probability of a throw
of an even number. Unless the theory makes it clear how alternatives
are to be selected, it is not clear what claim it is making about the meaning of probability statements. To cope with this difficulty, Laplace adopted what was subsequently called *the principle of indifference.* According
to this principle the alternatives have to be equiprobable, and two alternatives are equiprobable if we do not know any reason why one should
occur rather than the other. But the theory is still uninformative since
it does not answer the crucial and disputed question – what is a reason
why one alternative should occur rather than another?

As stated so far this theory does not look like a theory of epistemic
probability, probability on evidence. It seems to be concerned with some
probability which there is of an event occuring, which is quite independent of what we know of the circumstances. However it can be represented
as a theory of epistemic probability where the evidence is that certain
alternatives are (in the sense stated above) equiprobable. But in that case
it will not do as a general account of epistemic probability. Only rarely
when *e* renders *h* probable is our evidence *e* that there are certain alternatives, and we know no reason why one should occur rather than another.
When we calculate the probability of rain tomorrow from the position
of the barometer today and past correlations between rainfall and barometer readings, our evidence is not of this type. Nor is it when we cal-

culate the probability of a coin landing heads from the evidence that it
has landed heads 9/10 times previously. So Classical Theory will not do
as a general theory of epistemic probability, and, further, it provides no
way of recognizing those cases to which it does apply, cases where there
are a number of alternatives, and 'we do not know any reason' why one
should occur rather than another. We must look for some more general
account of epistemic probability, from which any limited truths of
Classical Theory can be derived.

The Subjective Theory was originally put forward by Ramsey and
subsequently developed by the statisticians de Finetti and Savage (see
the writings of these authors contained in [11]). On this view propositions about the probability of other propositions are propositions about
people's actual degree of belief in those other propositions. As an analysis
of propositions of epistemic probability this clearly fails. To say that it
is probable on such and such evidence available today that it will rain
tomorrow is not to say anything about how much anybody believes that
it will rain, for it could be probable whatever anyone believed. People
may not be thinking about whether it will rain or may assess the evidence
irrationally. The main motive behind the putting forward of the theory
has been not to provide an analysis of probability statements but to
provide an alternative meaning to the normal one of the 'probability'
referred to in the statements of the probability calculus, and so to provide
an alternative use for the calculus.

This leaves us with Logical Theory. The classical exposition of Logical
Theory is J. M. Keynes's *A Treatise on Probability* [12]. The theory has
been developed by many writers subsequently, including Carnap ([14],
[15] and [16]). On Logical Theory there are probability propositions
which state a relation which holds between evidence and hypothesis.
These propositions state how much certain evidence renders likely a certain hypothesis. Clearly, as I have illustrated earlier, there are such propositions, and I have termed them propositions of epistemic probability.
So we accept the claim of Logical Theory that there are propositions of
the kind it describes. Most of the advocates of Logical Theory, including
Carnap, have been prepared to admit the existence of other kinds of
probability propositions as well. (Carnap, [14] Chapter 2, distinguished
what he termed 'probability$_1$', that is epistemic probability, from what
he termed 'probability$_2$', that is statistical probability.)

Logical Theory makes three further claims. The first is that many
propositions which do not explicitly refer to evidence do so implicitly,
and that the evidence referred to implicitly is the total evidence available

at the time to which reference is made. To say that it is probable that Quantum Theory is true is to say that it is probable on the total evidence now available. To say that it is probable that Hitler committed suicide in 1945 is to say that it is probable on the total available evidence which we now have. This account seems correct. These statements concern the support which our current evidence gives to the propositions in question, that Quantum Theory is true, or that Hitler committed suicide in 1945. The total available evidence may be that available to some individual or to some community, and the context should make clear which is the individual or community whose evidence we are considering.

If this account is correct, it may seem surprising that we do not often use the phrase 'it was probable', saying e.g. 'it was probable two centuries ago that the world began in 4004 BC'. For if 'it is probable that *p*' is (often) elliptical for 'it is probable on the total evidence now available that *p*' we would expect 'it was probable that *p*' to be used as elliptical for 'it was probable on the total available evidence that *p*'. We may want to assess the extent to which the evidence we used rendered various hypotheses probable, but expressions of the form 'it was probable that *p*' sound odd. On the other hand we often say 'it *seemed* probable' that so and so, e.g. 'yesterday it seemed probable that he would come' or '3,000 years ago it seemed probable' that the earth was flat'.

I can only account for this fact of usage as follows. We are often interested more in how men of the past in fact assessed evidence than in the direction in which the evidence then pointed. Hence 'it seemed probable' will be used more than 'it was probable'. However we do sometimes say 'on the evidence available at the time it was probable that' so and so. We may say 'on the evidence available yesterday it was probable that there would be snow today' or 'on the evidence available last year it was improbable that prices would fall so quickly'. If we do not say 'it was probable' that so and so without the qualifying expression 'on the evidence then available', I can only conclude that this is a quirk of ordinary language without philosophical significance, and I commend the use of the phrase 'it was probable' as elliptical for 'it was probable on the (total) evidence available'.

The second claim which Logical Theory makes is that, given that we understand by the evidence all our knowledge relevant to assessing whether or not the hypothesis is true, the probability of the hypothesis on the evidence is independent of further empirical considerations. The truth or falsity of a proposition of epistemic probability is then a matter of non-empirical considerations alone. It is a matter of what are our criteria for assessing evidence. This is a claim for which I argued earlier (p. 12). But given that the correctness of our judgements of probability is a non-empirical matter, two possibilities remain. Either which propositions are probable on which evidence is something involved in what is meant by 'probable'; or it is a matter of standards taken for granted in our judgements about the world, but standards which do not merely record analytic truths. Logical Theory, by its very name, seems to suggest the former position – that our criteria for assessing evidence are involved in the very meanings of 'criteria', 'evidence', 'probability', etc., that statements ascribing probability are true of logical necessity, that is analytic. It has provided a well-known 'justification of induction' – that we are justified in using our rules for assessing probability, because no other rule for assessing 'probability' would be a rule for assessing 'probability' in the *normal* sense of the term. This kind of argument has been provided by Strawson [18].

The claim is however surely mistaken, and the most important advocate of Logical Theory, Carnap, certainly does not hold it. We can and do understand the concepts of 'probability', 'evidence' etc. from their relation to action and expectations. The claim that h is probable provides grounds for the claim that it is reasonable to act on h, to use it as a premiss in one's practical inferences etc., to wait for or expect h. (See pp. 198ff for further argument on this.) We would understand that speakers of another language were using a certain term '$\emptyset$' to mean 'probable' if in general when they agreed that h was '$\emptyset$', they took h for granted, expected h to happen etc. But they might have completely different grounds for saying that h was $\emptyset$ from those we have for saying that h is probable. We would naturally describe this as a situation where their grounds for assessing probability differed from ours. Hence the criteria we use for judging a proposition probable are ones which we intuitively judge to be reasonable, but it is logically possible that we adopt other criteria.[1] Despite its name and because of its history I will take Logical Theory to be compatible with this view.

The third claim which Logical Theory makes is that the probability of a hypothesis which ought, in absence of other considerations (see pp. 198ff. for these), to guide our actions, if we are to be rational, is that which is relative to the total evidence available to us. Thus the proposition h 'Eclipse will win the Derby' will have different probabilities on different pieces of evidence. e_1 may be that Eclipse did not win his last

(1) This claim has been developed in an article by Peter Winch 'Understanding a Primitive Society' *American Philosophical Quarterly*, 1964, 1, 307–24.

race, e_2 that Eclipse did not win his last race and that today the
ground is hard and Eclipse usually wins when the ground is hard.
$P(h/e_1)$ will not in general equal $P(h/e_2)$ which will not in general equal
$P(h/e_3)$ and so on. Let e_3 be the total evidence which we have and let
e_2 be part of it and e_1 part of e_2. Now in assessing the rationality of our
acting on h, the values of $P(h/e_1)$ and $P(h/e_2)$ are irrelevant – what matters
is $P(h/e_3)$. The greater is $P(h/e_3)$ the more rational it is to act on h, for
example to bet that h is true. This is clearly so, and is the claim of the
advocates of Logical Theory. But it has led to an important criticism by
A. J. Ayer [17]. Ayer points out that the propositions '$P(h/e_1) = p_1$',
'$P(h/e_2) = p_2$', '$P(h/e_3) = p_3$' etc. are all, according to Logical Theory,
necessary truths. But one necessary truth is as true as another. So what,
asks Ayer, makes it right for us to act on one rather than on another?

Ayer outlines ([17] p. 193) a possible answer to this objection. A
defender of Logical Theory might claim that to say that $P(h/e_3) = p_3$
entails that, if e_3 is our total available evidence, p_3 is that probability of
h which ought to guide our actions. The claim is that this principle of
total evidence is an analytic truth involved in the meaning of probability.
Considerations set forward above support this claim. But in that
case, Ayer objects, how can a defender of Logical Theory justify
our looking for more evidence? He points out that the sensible investigator of the truth of h looks for further evidence relevant for and against
the claim that h is true. The punter wants as much evidence as he can get
relevant to whether Eclipse will win the Derby, and considers his resultant
bet a more rational affair the more evidence on which it is based. But how
can Logical Theory account for this fact? If the obtaining of more evidence
means that our evidence is now e_4 instead of e_3 (e_4 including e_3), we will
now have a new probability relative to total available evidence, $P(h/e_4) = p_4$,
and this ought to guide our actions. But how can Logical Theory account
for it being a sign of rationality to look for more evidence, if all that ac-
quiring it does is to make it rational to act on one necessary truth rather
than another?

The answer to Ayer's question seems to be as follows. If we have to
decide whether or not to act on h, it is *not* always rational to look for
more evidence. If the truth or falsity of h is a comparatively unimpor-
tant matter or it would take a lot of time or trouble to get more evidence,
then it is not a mark of rationality to look for more evidence but it suffices
for us to be guided in our actions by that value of the probability, p_3,
which is relative to our present evidence e_3 ($P(h/e_3) = p_3$). Nor ought we
to look for more evidence (unless the truth or falsity of h is of quite

C

extraordinary importance) if $P(h/e_3)$ is very close to 1 or 0. If a telephone caller tells us that there is a stray cat in the basement (h) and we make a thorough search and cannot find it, and so on the evidence of the search (e_3) the probability of h is very low, extremely close to 0, it would be silly to search for more evidence. The only circumstances in which the rational man will search for more evidence are where the probability of the hypothesis is not very close to 1 or very close to 0 and where there is quite a possibility that the acquisition of more evidence may make a significant difference to the value of the probability. If a certain number of clues make it quite likely that Jones committed the murder, we ought to look for more clues because they may settle the matter (i.e. show beyond reasonable doubt that he did or that he didn't). We seek probabilities close to 1 or 0 for this reason, that we are less likely to make a mistake and so suffer undesired consequences if we act on hypotheses whose probability relative to the total evidence available to us is close to 1 rather than on other hypotheses, and less likely to make a mistake if we fail to act on hypotheses whose probability relative to the total evidence available to us is close to 0. I conclude that the rationality of searching for more evidence in certain circumstances is perfectly consistent with the Logical Theory of probability.

So we have adopted and expounded the Logical Theory of probability as an account of the meaning of a considerable number of propositions about probability. Such propositions state the extent to which certain evidence renders probable a certain hypothesis. They will either mention explicitly the propositions which form the evidence, or they will state that the total available evidence renders the hypothesis probable to such and such an extent, or they will merely state that the hypothesis is (or was) probable to such and such an extent. In the latter case if the proposition is a proposition of epistemic probability, then what it is saying is how probable the hypothesis is on the total evidence available at the time referred to.

Since this chapter has been much concerned with statistical probability, it will be appropriate to state in conclusion an important proposition governing the relations between epistemic and statistical probability. This is that (for values of q: $1/n$, $2/n$, $3/n \ldots n/n$):

$$P(\Phi_n\,(B/A) = q/Pr(B/A) = p \,.\, n\ A\text{'s are observed}) =$$

$$= \frac{n!}{(n - nq)!\ nq!}\ p^{nq}\ (1 - p)^{n\,-\,nq}.$$

To say that $Pr(B/A) = p$ is to say that, in a hypothetical population of
A's whose causal ancestors have occurred under initial conditions proportional to their actual occurrence and developed in the ways physically
most probable, a proportion p would be B. This hypothetical population
is large compared with any sample of A's studied. A sample of n A's is
now studied. We know nothing about this sample except that it is a sample from the hypothetical population. It seems to be a principle of epistemic probability that the epistemic probability that a given sample of
n A's has a certain proportion q of B's is equal to the proportion of
samples of n members from the population which have this proportion
of B's, given that all that is known about the sample is that it is an
n-fold sample from the population. Now it can be proved by simple
mathematics (I shall not give the proof here) that the proportion of
samples of n A's, nq of which are B's (for values of q: $1/n$, $2/n \ldots n/n$)
from a very large population of A's when the large population has
a proportion p of B's, is (to a high degree of approximation)

$$\frac{n!}{(n - nq)! \, nq!} \, p^{nq} \, (1 - p)^{n - nq} .$$

($n!$ is $1 \times 2 \times 3 \ldots \times n$; $(n - nq)!$ is
$1 \times 2 \times 3 \ldots \times (n - nq)$, and generally $z!$ - called factorial z - is all the
different integers up to and including z multiplied together.) Hence if
we ignore the approximation (as we can do, on the supposition that the
hypothetical population is very large), it follows that

$P(\Phi_n (B/A) = q/Pr(B/A) = p . n$ A's are observed) $=$

$$= \frac{n!}{(n - nq)! \, nq!} \, p^{nq} \, (1 - p)^{n - nq}.$$

From this principle two important propositions follow deductively.
One is that $P(Ba/Pr(B/A) = p . Aa) = p$. This principle says that on evidence that the complex statistical probability of an A being B is p and
that some individual object a is A, the epistemic probability of it being
B is p. (The deduction from the earlier principle is achieved by putting
$n = 1, q = 1$.)

The other proposition is that (for values of q: $1/n$, $2/n \ldots n/n$)
$P(\Phi_n(B/A) = q/Pr(B/A) = p . n$ A's are observed) $= r$ has its maximum
value when $q = p$ (or if p is not a possible value of q, q has the nearest
possible value to p) and gets smaller as q gets further away from p in either
direction (that is, for $q > p$, r gets smaller as $(q - p)$ gets larger; and for $q < p$,
r gets smaller as $(p - q)$ gets larger). To take an example, if our evidence
is that the (complex statistical) probability of a university teacher voting

Labour is 0·6 and we have a sample of 10 university teachers, then it is more probable that the number of Labour voters among them is 6 than that it is any other number. But it is more probable that there are 5 than that there are 4, more probable that there are 4 than that there are 3, more probable that there are 7 than that there are 8 and so on. (This can be seen fairly easily and proven by simple mathematics from the earlier principle, but I shall not give the proof here.)

We saw earlier in the chapter that if $\Pi(B/A) = p$ exists, $Pr(B/A)$ will also equal p. It follows that the above principles hold when 'Π' is substituted for 'Pr'.

Bibliography

For the Classical Theory of probability, see:

[1] P. S. DE LAPLACE *A Philosophical Essay on Probabilities* (first published 1814), English edition, New York, 1951.

For Frequency Theory, see:

[2] JOHN VENN *The Logic of Chance*, London, 1886.

[3] RICHARD VON MISES *Probability, Statistics, and Truth* (first published 1928), 2nd revised English edition, London, 1957.

[4] HANS REICHENBACH *The Theory of Probability* (first published 1934), Berkeley and Los Angeles, 1949.

[5] R. B. BRAITHWAITE *Scientific Explanation*, Cambridge, 1953, Chapters 5, 6 and 7.

[6] WESLEY SALMON *The Foundations of Scientific Inference*, Pittsburgh, 1966, Chapters 5 and 6.

For Propensity Theory, see:

[7] KARL R. POPPER 'The Propensity Interpretation of Probability', *British Journal for the Philosophy of Science*, 1960, **10**, 25–42.

[8] KARL R. POPPER 'Quantum Mechanics Without The Observer', especially pp. 28–34 in M. Bunge (ed.) *Quantum Theory and Reality*, New York, 1967.

[9] IAN HACKING *The Logic of Statistical Inference*, Cambridge, 1966.

[10] D. H. MELLOR *The Matter of Chance,* Cambridge, 1971.

For Subjective Theory, see:

[11] HENRY E. KYBURG and HOWARD E. SMOKLER (eds.) *Studies in Subjective Probability,* New York, 1964.

For Logical Theory, see:

[12] J. M. KEYNES *A Treatise on Probability,* London, 1921.

[13] W. C. KNEALE *Probability and Induction,* Oxford, 1949. (Kneale gives, as well as an account of epistemic probability in Parts I, II and IV of this book, an account of statistical probability in Part III.)

[14] RUDOLF CARNAP *The Logical Foundations of Probability* (first published 1950), 2nd edition, Chicago, 1962.

[15] RUDOLF CARNAP *The Continuum of Inductive Methods,* Chicago, 1952.

[16] RUDOLF CARNAP and RICHARD C. JEFFREY (eds.) *Studies in Inductive Logic and Probability. Volume I,* London, 1971.

For Ayer's criticism of Logical Theory, see:

[17] A. J. AYER 'The Conception of Probability as a Logical Relation' in *The Concept of a Person,* London, 1963.

For the claim that the rules of confirmation are analytic, see:

[18] P. F. STRAWSON *Introduction to Logical Theory,* London, 1952, Chapter 9, Part II.

CHAPTER III

The Probability Calculus

The axioms and theorems of the calculus of probability purport to state logical relations which propositions of probability bear to each other. Since the seventeenth century mathematicians have been calculating probabilities from other probabilities; and, as happened in other branches of mathematics such as geometry, later workers sought to axiomatize their knowledge of the field; that is to set up a small number of axions from which would follow deductively all the propositions about the logical relations between probabilities. The best known set of axions for the probability calculus is that due to Kolmogorov [2] in 1933, but similar sets have been put forward by others.

In Kolmogorov's system, as in most other systems to be found in books written by mathematicians, the axioms are expressed in set-theoretical terms, that is so as to state relations between sets or classes. '$Pr(\ /\ )$' and '$Pr(\ \ )$' are symbols denoting functions, the arguments of which denote classes. Classes are usually represented by capital letters A, B, C and so on. I will refer to members of A as A's, members of B as B's, and so on. '$Pr(B/A)$' denotes the probability of an A being B (which I will not for the moment interpret more specifically as in the last chapter as the complex statistical probability of an A being B).

I will now expound a typical set of such axioms of the calculus of probabilities. I take as a primitive symbol '$Pr(\ /\ )$'. In terms of this primitive symbol I define another symbol '$Pr(\ \ )$' as follows: Df: for all sets A, $Pr(A) \equiv Pr(A/V)$.

(V is the universal set, that is the set of all objects belonging to the universe of discourse.)

$Pr(A)$ is thus the probability of an object in the universe being A. Kolmogorov used '$Pr(\ \ )$' as the primitive symbol and defined '$Pr(\ /\ )$' by means of it in terms essentially of what I shall give as my Axiom IV. That is, for Kolmogorov, for all sets A and B '$Pr(A/B)$' is defined as

$\dfrac{\text{`}Pr(A \cap B)\text{'}}{Pr(B)}$ and so has a value only when $Pr(B) \neq 0$.[1] The advantage of my alternative procedure will be seen when we treat of the axioms in propositional instead of set-theoretical terms.

There are four axioms:

I For all sets A and B, $Pr(A/B) \geqslant 0$.

II For all sets A and B, if $B \subseteq A$ $Pr(A/B) = 1$.

III For all sets A, B and C if $A \cap B \cap C = 0$,
   $Pr(A \cup B/C) = Pr(A/C) + Pr(B/C)$.

IV For all sets A, B and C $Pr(A \cap B/C) = Pr(A/B \cap C) \times Pr(B/C)$.

If '$Pr(A/B)$' is understood in the way in which it was understood in the last chapter, and will be understood henceforward, the truth of the axioms will be apparent. The view there expounded is that propositions of complex statistical probability are propositions about the proportions of objects which have some property in a hypothetical population. $Pr(B/A) = p$ says that the proportion of objects which would be B in a hypothetical set of A's brought into existence under various initial conditions is p. Now the proportion of objects which have some property must be $\geqslant 0$. Hence Axiom I. If every B in a hypothetical set of B's is A, the proportion of B's which are A's is 1. Hence Axiom II. If there are n C's in the hypothetical population of which k are also A's, and j are B's, and no C is both A and B, then $(k + j)$ C's will be members of $(A \cup B)$. So since $\dfrac{k}{n} + \dfrac{j}{n} = \dfrac{k + j}{n}$ the proportion of C's which are members of $(A \cup B)$ will be the proportion of C's which are A plus the proportion of C's which are B. Hence Axiom III. If there are n C's of which j are B's, and m of those j are also A's, then the proportion of C's which are B is j/n, and the proportion of members of $(B \cap C)$ which are A's is m/j. In that case the proportion of C's which are members of $(A \cap B)$ is $\dfrac{m}{n} \cdot \left(\dfrac{m}{n} = \dfrac{m}{j} \times \dfrac{j}{n} \right)$. Hence Axiom IV.

Since the axioms hold for complex statistical probabilities, they will

(1) '$\cap$' is the intersection symbol. $A \cap B$ is the set of all objects which are both A and B. '$\cup$' is the union symbol. $A \cup B$ is the set of all objects which are either A or B or both. '$A \subseteq B$' says that the set of A's is included in the set of B's, that is that every object which is A is also B.

also hold for simple statistical probabilities. All the above arguments hold if the population is an actual one rather than a hypothetical one. Further, we showed in Chapter II that given that a physical probability $\Pi(B/A)$ exists it will have the same value as $Pr(B/A)$. Hence since the axioms hold for expressions of the form '$Pr(B/A)$', they also hold for such expressions when 'Π' is substituted for 'Pr', given that the physical probabilities so symbolized exist.

We now turn to axiomatizations of the probability calculus in which the axioms state relations between propositions. '$P(/)$' and '$P( )$' will replace '$Pr(/)$' and '$Pr( )$' and be used as symbols denoting functions which take as arguments propositions. Such an axiom system was first provided by Keynes [3], and the best known of such systems is due to Carnap [4]. In such an axiom system as in all axiom systems the terms may be understood in different ways, but in this book they will be understood in the usual way, which is that already outlined in Chapter I.[1] Thus '$P(p/q)$' will be understood as the epistemic probability of p, given q. The meaning of other expressions, apart from that to be given by definition of '$P( )$', has also been expounded in Chapter I, and they will be understood here and subsequently in that sense. The natural corresponding axioms in propositional terms to those given earlier in set-theoretical terms are the following:

DF   For all propositions p and q, $P(q) \equiv P(q/p \text{ v } \sim p)$.[2]
AXIOM I   For all propositions q and r, $P(q/r) \geqslant 0$.

---

(1) One other use to which the calculus has been put is as a system of many-valued logic. H. Reichenbach gave it this use in *The Theory of Probability*, Berkeley and Los Angeles, 1949, Chapter 10. For this use of the calculus see N. Rescher *Many Valued Logic*, New York, 1969, pp. 184–8.

(2) 'v' is the sign of disjunction. '$p \text{ v } q$' says that either p or q or both. '$p \text{ v } q$' is true if and only if either p is true or q is true, or both p and q are true; it is false if and only if both p is false and q is false. '$\supset$' is the sign of material implication. '$p \supset q$' is false if and only if p is true and q is false; otherwise it is true. '$p \supset q$' is often translated 'if p, then q' but it will be seen that this translation is misleading, for in ordinary language 'if p, then q' is not automatically true just because p is false, whereas '$p \supset q$' is automatically true if p is false. If '$p \supset q$' is true of logical necessity, I shall say that p entails q. '$\equiv$' is the sign of equivalence. '$p \equiv q$' says that both '$p \supset q$' and '$q \supset p$'. Hence '$p \equiv q$' is true if *either* both p and q are true *or* neither p nor q is true, but otherwise is false. Any proposition expressible with the aid of the propositional connectives '$\sim$', '$.$', 'v', '$\supset$' and '$\equiv$', which, given the rules for the interpretation of the connectives, cannot but be true, is called a tautology. '$(p . q) \supset (p \text{ v } q)$' is a tautology for any p and q. So is '$p \text{ v } (\sim p . q) \text{ v } (\sim p . \sim q)$'. And so is '$p \text{ v } \sim p$' for any p. A tautology is thus the simplest kind of analytic proposition.

AXIOM II* For all propositions q and r, if '$r \supset q$' is analytically true, $P(q/r) = 1$.

AXIOM III For all propositions p, q and r, if '$\sim(p . q . r .)$' is analytically true $P(p \vee q/r) = P(p/r) + P(q/r)$.

AXIOM IV For all propositions p, q and r $P(p . q/r) = P(p/q . r) \times P(q/r)$.

Axiom III is called the disjunctive (or addition) axiom; Axiom IV is called the conjunctive (or multiplication) axiom. I have marked the second axiom with an asterisk because it will be necessary to replace it by a weaker axiom, as we shall see shortly.

Since we are dealing with propositions instead of sets it seems necessary to add two further axioms. Sets are defined extensionally, that is by their members; and so two sets are the same sets if they have the same members. So if 'A' and 'B' are names of sets which have the same members ($A \subseteq B$ and $B \subseteq A$ and so $A = B$), they are names of the same set, and so can of course be substituted for each other. However propositions are not defined extensionally, that is by the possible worlds of which they are true; two propositions can be logically equivalent without being identical, that is without being the same proposition. Two propositions p and q are logically equivalent if whenever one is true, then of logical necessity the other must be (that is, '$p \equiv q$' is analytically true). So 'the moon is made of green cheese' is logically equivalent to 'the moon is made of green cheese, and $5 + 7 = 12$'. Yet these propositions, though logically equivalent, are not identical, that is they do not say the same thing. Hence if we suppose, as is plausible, that if r gives p a certain amount of probability, it gives q the same amount of probability if p and q are logically equivalent, we need an axiom to say so. This we will call Axiom V or the hypothesis equivalence condition. Similarly if we suppose that if p gives r a certain amount of probability, then q gives r the same amount of probability if p and q are logically equivalent, we need an axiom to say so. This we will call Axiom VI or the evidence equivalence condition.

AXIOM V For all propositions p, q, and r, if '$p \equiv q$' is analytically true $P(p/r) = P(q/r)$.

AXIOM VI For all propositions p, q, and r, if '$p \equiv q$' is analytically true $P(r/p) = P(r/q)$.

The advantages of treating '$P(/)$' as the primitive symbol in the axioms of epistemic probability, instead of defining it in terms of '$P( )$' by defining '$P(p/q)$' as $\dfrac{\text{'}P(p . q)\text{'}}{P(q)}$ are twofold. First, in all applications of the calculus $P(p/q)$ has a meaning which can be explained independently of the suggested definition. Secondly, $P(p/q)$ has a meaning which can be explained even if $P(q) = 0$, whereas on the suggested definition it would not be a significant expression if $P(q) = 0$. For these reasons I prefer to treat '$P(/)$' as an elementary symbol, and to define '$P( )$' in terms of it. The same procedure is followed by Carnap who provides a similar set of axioms for epistemic probability to mine. (See [4] especially pp. 284–92.) Carnap lays down my Axioms III, IV, V and VI, but has slightly different axioms from my I and II, although ones which yield for his special language the same deductive consequences as mine. However, my axiom set, as stated so far, contains a superfluous member, as Axiom V is a deductive consequence of Axioms II* and IV. I therefore substitute for my Axiom II* a weaker axiom:

AXIOM II   For all propositions p and q, $P(q/p \cdot (p \supset q)) = 1$.

The original Axiom II* can now be derived as a theorem from the new Axiom II and Axiom VI, as I shall show shortly.

One difficulty about these axioms of epistemic probability, raised in particular by Axioms III and IV, is that they seem to assume that all expressions $P(p/r)$ have exact quantitative, that is numerical, values. Axioms V and VI merely state that the probabilities of certain propositions are the same as each other, and are thus mere axioms of comparative probability. Axioms I and II make claims about probabilities being $\geqslant 0$, or being equal to 1. But to say that the probability of some proposition is 1 or that it is 0 is merely to say that it is certainly true or that it is certainly false, which is to say that the probability is maximal or that it is minimal, that no probability is greater than or less than this probability – which is again a matter of comparative probability. But Axioms III and IV tell us how to calculate for all p and r $P(p \vee q/r)$ and $P(p \cdot q/r)$ by processes of addition and multiplication on $P(p/r)$, etc; and, it might appear, addition and multiplication can only be performed on numbers. Hence it might seem that Axioms III and IV assume that $P(p/r)$ etc. always have exact quantitative values.

Yet clearly to the probabilities of some hypotheses we simply do not give exact quantitative values and there seems to be no obvious procedure for working out such values. Historians would agree that it is much more probable on our present evidence that Caesar crossed the Rubicon in 49 BC than that he crossed it in 39 BC. Our evidence fits the former hypothesis much better than the latter. Yet there seems no obvious procedure for working out the exact numerical probability of the former hypothesis or for working out how much more probable it is than the

latter one. Again although one can say that the General Theory of
Relativity is more probable on present evidence than many theories of
mechanics, it seems odd to say that its probability is $1/2$ or $2/3$ or $2/5$.
Now this oddness *may* arise because nobody has in fact worked out the
probability of the hypothesis from principles of probability implicit in
our understanding of the concept, although this is something which can
be done. That *may* be - but I do not know of principles which enable it
to be done, and intuitively the attempt to work out numerical probability
of such hypotheses as those quoted seems misguided. On the other hand
I do not know of a proof that such hypotheses do not have numerical
probabilities. The vast majority of writers do however make the latter
supposition. With them I shall make this supposition that *some* hypotheses do not have numerical probabilities although they can be compared in respect of probability with some other hypotheses.

It can be shown that if the axioms of the calculus govern probabilities
which do have quantitative values and all probabilities are commensurable,
then all probabilities have quantitative values. To say that all probabilities
are commensurable is to say that for any probability $P(q/r)$ and any other
probability $P(s/t)$, $P(q/r)$ is either greater than $P(s/t)$ or equal to $P(s/t)$
or less than $P(s/t)$. Now on the Logical Theory of epistemic probability
which we have adopted the value of a probability and so its relations of
greater than, etc., to other probabilities are necessary truths ascertainable
in principle by anyone who reflects on the criteria of epistemic probability which we use. This means that if $P(q/r) > P(s/t)$ this is something
demonstrable by someone who sits down to consider the matter. Now
imagine an urn in which there are n balls, m of which are red. By the
principles set out at the end of Chapter II, the probability of drawing a
red ball from a certain urn on the evidence that the urn contains n balls,
m of which are red is m/n. If all probabilities are commensurable, then
for any probability $P(h/e)$, and for any m and n, we can conclude whether
or not $P(h/e)$ is greater than, equals, or is less than the probability of
drawing a red ball from a certain urn on evidence that the urn contains
n balls, m of which are red. This means that for any probability $P(h/e)$
and for any m and n we can conclude whether $P(h/e)$ is greater than,
equals, or is less than m/n. In that case for any probability $P(h/e)$ we can
find *either* a value of m/n such that $P(h/e) = m/n$, in which case $P(h/e)$
equals a rational number, *or* distinct values of m/n as close as we like to
each other such that $P(h/e)$ lies between them, in which case $P(h/e)$ equals
a real number. Either way $P(h/e)$ has a quantitative value.

However the supposition that all probabilities are commensurable itself seems implausible. It seems implausible to suppose that anybody could find out whether the hypothesis that Caesar crossed the Rubicon in 49 BC is more probable on the current historical evidence than the hypothesis that the General Theory of Relativity is true on current scientific evidence. There seems to be no obvious procedure of finding out which of these hypotheses is more probable on the respective evidences. One can certainly often compare the probability of one hypothesis with that of another hypothesis on the same evidence (e.g. the probability of General Relativity with that of Newton's Theory on the same evidence) or of the same hypothesis on different evidences (e.g. General Relativity on current evidence rather than on last century's evidence). But there seems no reason for supposing that one can compare in respect of greater or less *any* probability $P(q/r)$ with any other $P(s/t)$. Hence it seems unjustifiable to assume that all probabilities are commensurable.

In the light of these considerations are we to say that the axioms of the calculus apply only to probabilities to which exact quantitative values can be attributed? We could say this and construct instead wider axioms, axioms solely of comparative probability, for all probabilities, from which we could derive axioms of quantitative probability for those propositions to which this can be ascribed. Such an axiom set has been constructed by Koopman [5]. However his axiom set is a complicated and unfamiliar one, and in order to keep the discussion within the context of the traditional axioms I shall, to meet the difficulties just outlined, adopt the following way of interpreting the latter axioms.

To say that the axioms of the calculus govern all probabilities of some kind will be to say that all axioms and theorems of the calculus are true of those probabilities which can be given quantitative values, and the axioms of comparative probability (viz. Axioms I and II, and V and VI) and theorems of comparative probability derivable from all six axioms of the calculus, are true of all probabilities. Another way of putting the point is as follows: to say that the axioms govern all probabilities of some kind is to say that if we give their true quantitative values to probabilities which have them, then we can give some quantitative values to probabilities which do not have true quantitative values, compatible with the true propositions of comparative probability which can be made about them, in such a way that the axioms will be true of those probabilities. In propositions of comparative probability the sign '$\geqslant$' will be interpreted as 'is not less than', not as 'is greater than or equal to'. Then an expression '$p_1 \geqslant p_2$' will not be taken to imply that p_1 and p_2 are commensurable probabilities. It says only that p_1 is not less than p_2, and that is compatible either with p_1 and p_2 not being commensurable or with p_1 having a value equal to p_2 or greater than p_2. Similarly '$\leqslant$' will be interpreted as 'is not greater than'.

So under the interpretation stated the axioms of the calculus are compatible with not all probabilities having quantitative values or being commensurable. Both of these claims – that not all probabilities have quantitative values or are commensurable – were put forward by Keynes ([3] Chapter 3) who, as stated, first set out a calculus of epistemic probability. It seems to me a very interesting synthetic fact about our criteria of probability that they are not precise enough to give us exact quantitative values for $P(h/e)$ for all h and e; or to say whether $P(h_1/e_1)$ is greater than $P(h_2/e_2)$, is equal to it, or is less than it, for all h_1, e_1, h_2 and e_2. There seems no *logical* reason why we should not have such precise measures, but in fact we do not.

I postulate, as do most confirmation theorists, that $P(q/r)$ exists for all q and r. In normal discourse we usually only talk about the probability of a proposition q where there is quite an amount of evidence concerned with the subject matter of q, but sometimes we talk about the probability of q on evidence only tenuously related to the subject matter of q. Confirmation theorists however usually also wish to talk about $P(q/r)$ where r is a mere tautology, and so by our definition $P(q/r)$ is $P(q)$. '$P(q)$' I will term the intrinsic probability of q. $P(q)$ is not a probability to which we refer in ordinary discourse. However our ordinary concept of probability is not such as to rule out its existence. The confirmation theorist is therefore not violating but somewhat extending the ordinary concept of probability in postulating the existence for all q and r of $P(q/r)$, and in particular in postulating the existence of intrinsic probabilities. His purpose in doing so is that intrinsic probability proves a useful concept for bringing out our ordinary understanding of probability.

We cannot ascertain the value of $P(q)$ by asking ourselves what we would say about the probability of q if we had no evidence except that of a tautology, for that is not a situation which we can imagine. However, there are various procedures which seem natural methods of inferring the value of $P(q)$, either a quantitative value or a comparative one (that $P(q)$ is greater or less than some other value), given that $P(q)$ exists. For example we shall show that the axioms of the calculus and other principles of epistemic probability derivable therefrom appear to show correctly how $P(q/r)$ is related to other epistemic probabilities in all cases when we can make intuitive judgements, whatever the evidence r is in $P(q/r)$. On these grounds it seems reasonable to postulate that these principles hold without exception, from which it follows that they hold when r is a tautology, i.e. when $P(q/r)$ is $P(q)$. This means that we can often use the principles to calculate a quantitative or comparative value for $P(q)$. The value of $P(q)$ like the values of all epistemic probabilities is a non-empirical matter, and depends solely on what 'q' says. Once we can ascertain the values of some intrinsic probabilities, we can inspect our results and see what features of q determine $P(q)$. (Chapter VII is largely an exercise in doing just this.) Once we have ascertained this, we can work out the values of other intrinsic probabilities. Of course if the various methods for ascertaining the value of $P(q)$ yielded different results that would be a reason for denying the uniqueness or existence of $P(q)$; but I know of no reason for supposing that this happens.

With the qualifications stated above on their interpretation can we justify the axioms of the probability calculus expressed in propositional terms as axioms of epistemic probability? Do they state logically necessary relations between propositions of epistemic probability? Axioms I and II clearly do this. '$P(q/r) = 0$' says that on evidence r, $\sim q$ is certain or it is certain that q is false. No evidence can render a proposition less certain than that. Hence Axiom I. If we know that p is true and we know that q is true when p is true, then surely we know that q is true; q is as certain as our evidence. Hence Axiom II. However, whether the remaining axioms hold of all propositions of epistemic probability has been disputed. I will discuss in the next chapter whether Axioms V and VI hold and in Chapter V whether Axioms III and IV hold.

I will now prove some of the important theorems which follow from the axioms of the calculus expressed in propositional terms. (Corresponding theorems can be proved in the calculus expressed in set-theoretical terms.) The theorems to be stated hold whatever propositions 'p', 'q', 'r' etc. denote.

Axiom II says $P(q/p \, . \, (p \supset q)) = 1$. If $p \supset q$ is analytically true, '$p \, . \, (p \supset q)$' is logically equivalent to 'p'. Hence, using Axiom VI, we get:

Theorem 1 If '$p \supset q$' is analytically true $P(q/p) = 1$. (This was our original second axiom – Axiom II*.)

By this theorem the probability of any analytic proposition r on any evidence e will be 1, since '$e \supset r$' will be analytically true.

By Axiom III since '$\sim(q \, . \sim q \, . \, r)$' is a tautology and so analytically true, $P(q \lor \sim q/r) = P(q/r) + P(\sim q/r)$. Since '$q \lor \sim q$' is a tautology and so analytically true, by Theorem 1, $P(q \lor \sim q/r) = 1$ and so

Theorem 2 $P(q/r) + P(\sim q/r) = 1$.

By Axiom I both $P(q/r)$ and $P(\sim q/r) \geqslant 0$. Hence

   Theorem 3 $P(q/r) \leqslant 1$.

   By Axiom III $P(p \vee (\sim p . q)/r)$ $= P(p/r) + P(\sim p . q/r)$.

   By Axiom IV $P(\sim p . q/r)$ $= P(\sim p/q . r) \times P(q/r)$.

   Hence by Theorem 2 $P(\sim p . q/r) = \{ 1 - P(p/q . r) \} \times P(q/r)$

                                      $= P(q/r) - P(p/q . r) \times P(q/r)$.

   By Axiom IV $= P(q/r) - P(p . q/r)$.

   Hence $P(p \vee (\sim p . q)/r)$ $= P(p/r) + P(q/r) - P(p . q/r)$.

   From there, using Axiom V, we get

   Theorem 4 $P(p \vee q/r)$ $= P(p/r) + P(q/r) - P(p . q/r)$.

This addition theorem is more general than the addition Axiom III which applies only when '$\sim(p . q . r)$' is analytically true. Theorem 4 applies whatever the logical relations between p, q and r.

A useful definition is the definition of independence or irrelevance. Given r, q is independent of p(or - see the definition given on p. 4 - given r, p is irrelevant to q) if and only if $P(q/r) = P(q/p . r)$. On normal information (r) about the form of the players involved, the proposition (q) that Oxford wins the boat race is independent of the proposition (p) that Tal wins the world chess championship, since the addition of p to r does not affect the probability of q. The information that Tal has won the world chess championship would not in general increase or decrease the probability of Oxford winning the boat race. Of course there could be evidence r on which the cited q was not independent of the cited p - e.g. evidence that a gambling syndicate had made one of two decisions - either to bribe both Tal and the members of the Oxford crew to lose their respective competitions, or to bribe both of their opponents to lose their respective competitions. But on normal sporting information r, the cited q is independent of the cited p.

Axiom IV states that $P(p . q/r) = P(p/q . r) \times P(q/r)$. Hence

   Theorem 5 If p is independent of q, given r

      $P(p . q/r) = P(p/r) \times P(q/r)$.

It follows from Axiom IV by substituting 'p' for 'q', and 'q' for 'p' that $P(p . q/r) = P(q/p . r) \times P(p/r)$. Hence if p is independent of q, given r,

      $P(p . q/r) = P(p/r) \times P(q/r) = P(q/p . r) \times P(p/r)$.

Hence, unless $P(p/r) = 0$, $P(q/r) = P(q/p . r)$, which is to say that q is independent of p, given r. Hence

   Theorem 6 p is independent of q, given r, if q is independent of p, given r (given that $P(p/r) \neq 0$).

Hence (given that $P(p/r) \neq 0$ and $P(q/r) \neq 0$) we talk of p and q being independent of each other given r, if one is independent of the other, given r.

Since by Axiom IV

$$P(e \cdot h/k) = P(h/k) \times P(e/h \cdot k) = P(h/e \cdot k) \times P(e/k)$$

there follows

Bayes's Theorem If $P(e/k) \neq 0, P(h/e \cdot k) = \dfrac{P(h/k) \times P(e/h \cdot k)}{P(e/k)}$.

Where h is the hypothesis whose probability we are assessing, k our background evidence and e the new evidence, then '$P(h/k)$' is called the prior probability of h in contrast to '$P(h/k \cdot e)$', the posterior probability of h. Where k is a tautology, $P(h/k)$ is $P(h)$, the intrinsic probability of h. '$P(e/h \cdot k)$' I shall term a measure of the accuracy of h. It is a measure of how much the new evidence e is to be expected, for given background evidence k, if the hypothesis h is true. The statistician R. A. Fisher introduced the use of the term 'likelihood' in this connexion, and said that in so far as $P(e/h_1 \cdot k) > P(e/h_2 \cdot k)$, h_1 had greater likelihood than h_2. (See R. A. Fisher *Statistical Methods and Scientific Inference*, 2nd edition, Edinburgh, 1959, pp. 68ff.) I shall however avoid this use, as I find it less confusing to use 'likelihood' in a non-technical sense, and say instead that in so far as $P(e/h_1 \cdot k) > P(e/h_2 \cdot k)$, h_1 is more accurate than h_2.

There follow from the axioms of the calculus a number of further theorems which are of great importance for our topic and which have been advocated by, among others, writers who do not hold that the axioms of the calculus of epistemic probability are in general true. I shall now state these, and produce, as those writers do, independent justification of them. In order to emphasize that independent justification can be given, I shall call them principles of epistemic probability, and label them by letters A to E rather than by numbers.

Each principle is preceded by a qualification to the effect that the principle holds subject to certain values (e.g. $P(e/k)$) not being equal to 0. The principles only follow from the axioms of the calculus given such a qualification in each case – although some writers have wished to claim for some of the principles that they hold without a qualification. I shall not however make the latter supposition. To say that a probability is not equal to 0 is to say (in view of Axiom I) that it is greater than 0, which is to say that it is greater than that of a proposition which is certainly false, e.g. a proposition incompatible with the evidence. We might be inclined to suppose that $P(p/r)$ is always greater than 0 if p is compatible with r. For after all, given r there is in that case some possibility that p, whereas there is no possibility that $\sim r$, so it would appear that $P(p/r) > P(\sim r/r)$. $P(\sim r/r) = 0$, hence $P(p/r) > 0$. However we shall meet in Chapter V a very serious objection to the claim that all propositions compatible with their evidence have probabilities greater than 0. However, the counter-instances to be considered in Chapter V concern cases of propositions p which could never be observed to hold and so form part of our actual evidence. There seems no objection to the following weaker claim, rendered plausible by considerations given above. This is that any proposition p which we do not know to be false (i.e. is compatible with our present evidence) and could one day know to be true (i.e. could come to form part of our actual evidence) has a probability on our present evidence greater than 0. (When I write that 'p' could come to form part of our actual evidence, I do not mean that 'p' could occupy the evidence place in a probability function '$P(\ /\ )$', for any proposition can do that. We can always, for any r, assess the value of $P(r/p)$; in assessing the value of $P(r/p)$, we are assessing how probable r would be *if p were* our evidence. By 'p could come to form part of our actual evidence' I mean that it is logically possible that we could be in a position where p is part of our evidence.) A proposition which could come to form part of our evidence I will term an observation report. I shall not offer a general philosophical theory about what sorts of propositions are observation reports, but I shall in the course of argument refer to various propositions which clearly are and others which clearly are not observation reports.

Hence if we are assessing the probability of h when k is our background evidence and e our new evidence $P(e/k)$ will never equal 0. For if we know both e and k they must be compatible, and clearly e is the kind of proposition which can serve as evidence. For the latter reason $P(e/h.k)$ will equal 0 only if e is incompatible with $(h.k)$. $P(h/e.k)$ and $P(h/k)$ will equal 0 if h is incompatible with $(e.k)$ and k respectively; but we will allow for the possibility that these probabilities may equal 0 under other circumstances to be discussed in Chapter V.

With these preliminaries, let us set out the principles. Principles A to D follow immediately from Bayes's Theorem.

PRINCIPLE A (Given that $P(e/k) \neq 0$ and $P(e/h_1.k) \neq 0$) if $P(e/h_1.k) = P(e/h_2.k)$, then $P(h_1/e.k) > P(h_2/e.k)$ if and only if $P(h_1/k) > P(h_2/k)$; and if $P(e/h_1.k) > P(e/h_2.k)$, then $P(h_1/e.k) > P(h_2/e.k)$ if $P(h_1/k) > P(h_2/k)$.

If for the moment we ignore the qualification, then the first part of the principle says that given two hypotheses h_1 and h_2 which are equally

D

accurate in their prediction of new evidence, then h_1 has the greater posterior probability if and only if it has the greater prior probability. This seems intuitively obvious. Suppose that h_1 and h_2 are two theories about how a murder was done. Some new evidence e turns up which is equally much or little to be expected on either hypothesis ($P(e/h_1 . k) = P(e/h_2 . k)$). If h_1 is more probable than h_2 after the discovery of e, it can only be because it was more probable than h_2 before the discovery; and if h_1 was more probable than h_2 before the discovery, it will in these circumstances remain so after it. The second part of the principle says, if we continue to ignore the qualification, that if h_1 is more accurate in predicting e than is h_2 *and* was more probable than h_2 before e was discovered, it remains more probable after the discovery. If the first part of the principle holds, then *a fortiori* the second part holds.

The principle is subject to the qualification 'given that $P(e/k) \neq 0$ and $P(e/h_1 . k) \neq 0$'. Now I suggested on pp. 43f. that if e is an observation report (i.e. something which can come to form part of our evidence) then $P(e/k)$ can only equal 0 if e is inconsistent with k, and $P(e/h_1 . k)$ can only equal 0 if e is inconsistent with $(h_1 . k)$. Now if e and k are both parts of our evidence $P(e/k)$ cannot equal 0, for we cannot know both of two inconsistent propositions. If $P(e/h_1 . k)$ does equal 0, because e is inconsistent with $(h_1 . k)$, the claim of Principle A will certainly not hold. For even if h_1 has greater prior probability than h_2, it does not have greater posterior probability than h_2 if it is incompatible with the new evidence e. In that case its probability will be 0.

Having provided justification of Principle A for cases where we can assess intuitively all the values $P(h_1/k), P(e/h_1 . k)$, and so on involved, let us suppose the principle universally applicable, and use it to show something about intrinsic probability. We will do this by introducing what is known as the curve-fitting problem. Imagine a scientist studying the relation between the pressure and volume of a gas at constant temperature, who knows nothing about the behaviour of other gases, but has as his total evidence (e) that pressure was read at certain (approximate) values of the volume (e_2) and that certain approximate values of the pressure were found for each (e_1), together with other knowledge about the world (e_3) of a type which is a non-technical sense we would say to be irrelevant to hypotheses about gas behaviour (e_3) and also a tautology k. 'e' is '$e_1 . e_2 . e_3 . k$'. The scientist seeks for the equation governing the relation of pressure and volume. Let the points marked with a cross in Figure 1 represent values of the pressure found for observed values of volume. The scientist seeks a curve which passes through those points.

The curve will show for each value of the volume what value (at any rate to a high degree of approximation) of the pressure will be found. It can be expressed as a theory consisting of one nomological proposition of the form 'of physical necessity all values v of the volume and p of the pressure are values of which $p = jf(v)$ is true'. $f(v)$ is some function of v, such as v^2 or v^3 or $1/v$, such that $p = f(v)$ is an equation which gives a unique value of the pressure for each value of the volume. j is either '1' or '1 ± r' where r is some small constant such as 0·05. In the former case '$p = jf(v)$' gives a unique value of the pressure for each value of the volume;

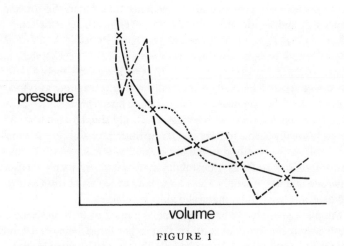

FIGURE 1

in the latter case '$p = jf(v)$' gives a value of the pressure for each value of the volume within a limit of error. In Chapter V I shall give reasons for supposing that in general confirmation can only be obtained by hypotheses of the latter type, that is hypotheses in which constants are given only an approximate value.

Now for all theories T, which, given e_2 and e_3, predict e_1 ($P(e_1/T . e_2 . e_3)$ = 1), e_2 and e_3 are equally likely to be found. Hence the total evidence e is equally probable on any such theory (given k). ($P(e/T . k)$ is the same for all such theories.) These theories are represented by curves which pass through the points marked on the graph. (Three such curves are marked on the graph above.) Yet the number of curves which do this will be infinite, and so too therefore will be the number of mutually incompatible nomological propositions compatible with e, stating what is the relation

between pressure and volume. This remains so even if we confine ourselves to nomological propositions which yield only an approximate value
of the pressure for a given value of the volume. For since the volume can
take any of an infinite number of values, an infinite number of different
nomological propositions can be constructed which differ from a given
one in making a different prediction about pressure from it for one value
of the volume (apart from the finite number of values of the volume at
which pressure was measured, the value of which volumes we know
approximately).

Clearly however we regard some of the theories T for which $P(e/T . k)$
is the same as more probable on the evidence than others. So for some
theories T_1 and T_2 for which $P(e/T_1 . k) = P(e/T_2 . k)$, nevertheless
$P(T_1/e . k) > P(T_2/e . k)$. This can only be, given Principle A, if $P(T_1/k) >$
$P(T_2/k)$, and k being a tautology this means if $P(T_1) > P(T_2)$, if T_1 has
greater intrinsic probability than T_2. Thus we regard the theory claiming
that $p = $ constant$/v$ marked on the figure with a continuous curve as more
probable than the one marked with a dashed line. This seems to be because it can be described with a mathematically simpler equation. A
theory normally seems to have greater intrinsic probability, the simpler
it is. I follow the tradition in the philosophy of science which terms a
theory simple in so far as it postulates mathematically simple relations
between variables, has only a few laws, laws which fit neatly together,
and does not have odd unnatural qualifying clauses.

Simplicity has always been regarded as a mark of truth in science. If
one theory is simpler than another, that, other things being equal, makes
it more likely to be true. Thus Kepler's Theory of Planetary Motion had
only three nomological propositions:

1 All planets describe ellipses with the sun at one focus;

2 All planets so move relative to the sun that the line joining a planet
   and the sun sweeps out equal areas in equal times;

3 All planets so move that the square of their period of revolution
   round the sun is proportional to the cube of their mean distance
   from the sun.

The fact that this theory consisted of so few nomological propositions,
short and comprehensible, has always been seen as showing it to be more
likely to be true than theories such as those which preceded it, consisting
of very many laws, several separate ones for each planet.

We shall examine at length in Chapter VII what makes one theory intrinsically more probable than another, and consider more fully this notion of simplicity which is so important in determining intrinsic probability.

There also follows directly from Bayes's Theorem:

PRINCIPLE B (Given that $P(e/k) \neq 0$ and $P(h_1/k) \neq 0$) if $P(h_1/k)$
$= P(h_2/k)$, then $P(h_1/e \cdot k) > P(h_2/e \cdot k)$ if and only if
$P(e/h_1 \cdot k) > P(e/h_2 \cdot k)$.

In words, if we ignore the qualification, Principle B says that for two hypotheses h_1 and h_2 with equal prior probability the one which is more accurate in its predictions has greater posterior probability. Let us consider, as before, two theories h_1 and h_2 about how a murder was done, and now suppose that on the evidence k available before the discovery of e, there is nothing to choose between them – each is as likely as the other. Now suppose new evidence e turns up which is just what is to be expected given h_1, but very puzzling if h_2 is true. h_1 might be the claim that Jones did the murder and that Smith is completely innocent; h_2 be the claim that Smith did the murder and that Jones is completely innocent; and e be the discovery of Jones's fingerprints on the murder weapon. Then e is quite likely if h_1 is true, but very unlikely if h_2 is true (both given k which, let us suppose, includes knowledge of how fingerprints are produced) – $P(e/h_1 \cdot k) > P(e/h_2 \cdot k)$. Then we would say that after the discovery of e, h_1 was more probable than h_2. Conversely if h_1 and h_2 are equally likely before the discovery of some new evidence e, but h_1 more likely after the discovery, it can only be because the new evidence was more to be expected on h_1.

If we take k as a mere tautology, then Principle B tells us that if we have two theories of equal intrinsic probability, the more accurate one is the one is more likely to be true.

Principle B is subject to the qualification 'given that $P(e/k) \neq 0$ and $P(h_1/k) \neq 0$'. If e is new evidence added to k, then, for the reason given on p. 43, $P(e/k) \neq 0$. $P(h_1/k)$ will equal 0 if h_1 is incompatible with k, but in that case $P(h_1/e \cdot k)$ will also equal 0, since h_1 will be incompatible with $(e \cdot k)$ and so $P(h_1/e \cdot k)$ cannot possibly exceed another probability value. However there is the possibility, to be discussed in Chapter V, that $P(h_1/k)$ may equal 0 without k being incompatible with h_1. I shall not attempt to provide an independent justification for this restriction 'given $P(h_1/k) \neq 0$' imposed on Principle B for such a case, but merely point out that the principle follows from the axioms of the calculus only with that restriction.

Our next principle to follow directly from Bayes's Theorem is:

PRINCIPLE C (Given that neither $P(h/k)$ nor $P(e/k)$ equals 0) e confirms h in relation to background knowledge k, that is $P(h/e \cdot k) > P(h/k)$ if and only if $P(e/h \cdot k) > P(e/k)$, and disconfirms h in relation to k, that is $P(h/e \cdot k) < P(h/k)$ if and only if $P(e/k \cdot h) < P(e/k)$.   misprint

New evidence, that is, confirms a hypothesis if and only if that evidence is more probable given the hypothesis and background evidence than it is given background evidence alone. If we omit the qualification 'given that neither $P(h/k)$ nor $P(e/k)$ equals 0', Principle C is the basic part (the part which he called 'C') of a principle advocated by J. L. Mackie [6] and called by him 'the relevance criterion'. Principle C is evidently true. New evidence e only confirms, supports, makes more likely to be true than it was, a hypothesis h if it is evidence which the knowledge we already have would not lead us to expect as much as it would if h were joined to it. Robinson's fingerprints on the safe only confirm the hypothesis that he robbed the safe if the fingerprints were not as likely to be found given only our other evidence. If Robinson is known to be the manager of the shop in which the safe is situated, and so his fingerprints are to be expected on the safe, whether or not he robbed it, the fact that his fingerprints are found there does not confirm the hypothesis that he robbed it.

$P(e/k)$ will not equal 0 if e is our new evidence added to k for the reason given earlier. $P(h/k)$ will equal 0 if h is incompatible with k and in that case of course e cannot possibly confirm h. But, as I shall show in Chapter V, $P(h/k)$ may equal 0 even if h is compatible with k. As with the similar restriction on Principle B I shall not attempt yet to justify the restriction on Principle C imposed by 'given $P(h/k) \neq 0$' for such a case; merely point out that Principle C follows from the axioms of the calculus only with this restriction.

It follows from Principle C that a universal nomological hypothesis, such as 'of physical necessity all A's are Q' (h), is confirmed by discovering of a known A that it is Q, given that $P(h/k) \neq 0$. Let k include evidence that an object a is A, but not include evidence that a is Q, and e be that a is Q. Since it can be discovered and so known that a is Q, by the arguments of p. 43 $P(e/k) > 0$, and since k did not include the knowledge that a was Q $P(e/k) < 1$. $P(e/h \cdot k) = 1$. Hence $P(e/h \cdot k) > P(e/k)$. $P(e/k) \neq 0$. So if $P(h/k) > 0$, $P(h/e \cdot k) > P(h/k)$, which is to say that h is confirmed by e (for background evidence k). But given Bayes's Theorem (p. 42) it will be seen that in this case if $P(h/k) = 0$, h can never be confirmed by any observation report e, that is any evidence for which $P(e/k) > 0$. For any such evidence e, $P(h/e . k) = 0$, and so $P(h/e . k) = P(h/k)$.

There follows from Bayes's Theorem the further principle:

PRINCIPLE D (given that neither $P(h/k)$ nor $P(e_1/k)$ nor $P(e_2/k)$ equals 0) e_1 confirms h better than e_2 does (which is to say – see p. 4 – $\dfrac{P(h/e_1 . k)}{P(h/k)} > \dfrac{P(h/e_2 . k)}{P(h/k)}$) if and only if $\dfrac{P(e_1/k . h)}{P(e_1/k)} > \dfrac{P(e_2/k . h)}{P(e_2/k)}$.

Principle D is virtually the same as another part (termed by Mackie 'C₁') of Mackie's relevance criterion. In words, the principle says that (for background knowledge k) if e_1 and e_2 both confirm h_1, e_1 confirms it more than e_2 does if and only if the addition of h to k raises the probability of e_1 by more than it raises the probability of e_2; if e_1 disconfirms h it disconfirms it less than e_2 does if and only if the addition of h to k lowers the probability of e_1 by less than it lowers the probability of e_2.

The principle is plausible for reasons given by Mackie in [6] (I substitute my 'e' for Mackie's 'b'):

Suppose that h is, or would be, confirmed by each of two observations e_1 and e_2 independently, and suppose further that $k . h$ entails both e_1 and e_2, but that $P(e_1/k)$ is less than $P(e_2/k)$. In other words if the hypothesis is true each of the observations will be made when the corresponding experiment is performed, but apart from the hypothesis, on the basis only of our other beliefs, a favourable result from the first experiment is much less likely than a favourable result from the second. Then the criterion C_1 states that the actual observation e_1 would confirm h better than the actual observation e_2. Consider, as an example, the hypothesis that a certain quantity of a certain drug taken every twenty-four hours gives complete immunity to malaria. In one test, ten men take the drug and live for three months in a highly malarious area; in another test, one man takes it and lives for three months in a slightly malarious area. Here $k . h$ entails both e_1 (none of the ten will get malaria) and e_2 (the one man will not get malaria). But e_1, that is a favourable result of the first test, will surely confirm h much better than e_2, a favourable result of the second. The first is a more thorough test. And why? Surely because the favourable result of the first was intrinsically less likely, and therefore, since $P(e_1/k \cdot h) = P(e_2/k \cdot h) = 1$, there was a bigger gap between $P(e_1/k \cdot h)$ and $P(e_1/k)$ than between $P(e_2/k \cdot h)$ and $P(e_2/k)$.[1]

Principle D is subject to the restriction 'given that neither $P(h/k)$ nor $P(e_1/k)$ nor $P(e_2/k)$ equals 0'. Given that e_1 and e_2 are pieces of evidence which can be added to our background evidence k, $P(e_1/k)$ and $P(e_2/k)$ will not equal 0. If $P(h/k) = 0$, the Principle obviously does not hold because $\dfrac{P(h/e_1 \cdot k)}{P(h/k)}$ and $\dfrac{P(h/e_2 \cdot k)}{P(h/k)}$ are not significant expressions, since division by 0 is not an operation recognized in mathematics.

An important result which follows from Principle D is that a number of observations predicted by some hypothesis h for given background evidence k confirm the hypothesis better if they are observed under a variety of conditions under which the phenomena of the kind in question usually vary if they do vary. Let h be the hypothesis 'all swans are white' and k the background evidence that thousands of swans have been observed in all regions of the world, including a hundred in England, and that other species of birds if they have varieties of different colours usually have them in different geographical regions (e.g. that if some species of bird has a variety which is yellow and a variety which is blue, then the yellow birds and the blue birds are usually found in different geographical regions). Let e_2 be that all of a hundred swans observed in England are white, and e_1 be that all of a hundred swans observed in various regions throughout the world are white. $P(e_1/k \cdot h) = P(e_2/k \cdot h) = 1$. The hypothesis that all swans are white together with evidence of the existence of swans in various places entails that all swans observed are white. But $P(e_2/k) > P(e_1/k)$. On evidence k one is more likely to find that all of a hundred swans observed in one region are white than that all of a hundred swans observed in different regions of the world are white. This is because the former is quite likely to be the case, our

(1) In [6] Mackie put forward three parts of his relevance criterion – 'C', which is virtually the same as our Principle C, and 'C$_1$', which is virtually the same as our Principle D (both of which I have defended) and also a Principle 'C$_2$'. 'C$_2$' is also a deductive consequence of Bayes's Theorem – given that $P(h_1/k) \neq 0$, $P(h_2/k) \neq 0$ and $P(e/k) \neq 0$. (If any of these terms equals 0 the expression in which it occurs as a divisor has no value in mathematics.) 'C$_2$' states that, given k, e confirms h_1 better than it does h_2 $\left(\dfrac{P(h_1/e \cdot k)}{P(h_1/k)} > \dfrac{P(h_2/e \cdot k)}{P(h_2/k)} \right)$ if and only if $\dfrac{P(e/k \cdot h_1)}{P(e/k)} > \dfrac{P(e/k \cdot h_2)}{P(e/k)}$. An argument against this by Schlesinger will be considered in Chapter VI.

evidence tells us, even if there are many swans of non-white varieties, whereas the latter is not very likely to be the case if there are many swans of such varieties. For if there are non-white swans examination of different regions of the world is more likely to detect them than the same amount of examination in only one. Hence $P(e_2/k) > P(e_1/k)$. Therefore $\dfrac{P(e_1/k \cdot h)}{P(e_1/k)} > \dfrac{P(e_2/k \cdot h)}{P(e_2/k)}$ and so by Principle D e_1 confirms h more than e_2 does – a result which surely intuitively we feel to be correct.

By Theorem 1 the probability of any analytic proposition on any evidence is 1. Let '$h_1 \supset h_2$' and so '$\sim h_1 \vee h_2$' be analytically true. Then for all e $P(\sim h_1 \vee h_2/e) = 1$.

By Theorem 4, $P(\sim h_1 \vee h_2/e) = P(\sim h_1/e) + P(h_2/e) - P(\sim h_1 \cdot h_2/e)$.

By Theorem 2, $P(\sim h_1/e) = 1 - P(h_1/e)$.

Hence $P(h_2/e) = P(h_1/e) + P(\sim h_1 \cdot h_2/e)$.

By Axiom I, $P(\sim h_1 \cdot h_2/e) \geqslant 0$. Hence:

PRINCIPLE E   If '$h_1 \supset h_2$' is analytically true, $P(h_2/e) \geqslant P(h_1/e)$ and unless $P(\sim h_1 \cdot h_2/e) = 0$, $P(h_2/e) > P(h_1/e)$.

This means that on any evidence part of a theory is no less likely to be true as the whole theory; the prior and posterior probabilities of the latter will be no greater than those of the former; and unless there is no probability at all of the part being true without the whole being true, then the probability of the part will be less than that of the whole. Again I suggest that this principle is intuitively plausible. The more you say, the more likely you are to make at least one mistake. If you claim that h_2 then there is a certain probability (dependent on the evidence) of your being wrong. But if you make a further claim which has some probability of being false then you add to the probability of your making at least one mistake. If '$h_1 \supset h_2$' is analytically true, by claiming 'h_1' you claim at least that h_2, and so are at least as likely to be wrong as if you claimed merely that h_2. But if $P(\sim h_1 \cdot h_2/e) \neq 0$ and so by claiming that h_1 you make a further claim than that h_2, which has some probability of being false, you add to the probability of your being wrong.

In this Chapter I have set out the axioms of the probability calculus and derived from them important theorems and principles. I have provided justification of the first two axioms of the calculus and also of important derived principles, so that even if some of the arguments to be considered against the other axioms of the calculus have force, each of these principles has some independent justification. However we shall

discuss in Chapter VI a number of arguments directed against the claim
that the axioms of the probability calculus are true, some of which are
also arguments against the five principles stated in this chapter for which
we have produced such independent justification.

Bibliography

For an elementary mathematical treatment of the calculus of probability
in set-theoretical terms, see:

[1] HARALD CRAMER *The Elements of Probability Theory*, New York,
     1955.

For Kolmogorov's axioms of the probability calculus in set-theoretical
terms, see:

[2] A. N. KOLMOGOROV *Foundations of the Theory of Probability*
     (first published 1933), English edition, New York, 1950.

For Keynes's and Carnap's axioms and theorems of the probability calculus in propositional terms, see:

[3] J. M. KEYNES *A Treatise on Probability*, London, 1921.

[4] RUDOLF CARNAP *Logical Foundations of Probability* (first pub-
     lished 1950), 2nd edition, Chicago, 1962.

For Koopman's axioms of comparative probability, see:

[5] B. O. KOOPMAN 'The axioms and algebra of intuitive probability'
     Annals of Mathematics, 1940, **41**, 269–92.

For discussion of Principles C and D, see:

[6] J. L. MACKIE 'The Relevance Criterion of Confirmation' *British
     Journal for the Philosophy of Science*, 1969, **20**, 27–40.

CHAPTER IV

The Equivalence Conditions

Having provided in the last chapter justification of the first two axioms of the calculus, as well as of some derived principles, I propose in this chapter to examine Axioms V and VI, the two equivalence conditions. Let us begin with the hypothesis equivalence condition:

AXIOM V   For all propositions p, q, and r, if '$p \equiv q$' is analytically true $P(p/r) = P(q/r)$.

Superficially this condition is plausible. If r renders p to some extent likely to be true, it renders likely to that extent the claim that a certain state of the universe described by p holds. That state is the state described by q. Yet the support given by the evidence to the existence of that state is surely independent of how the state is described.

Yet there are arguments against the hypothesis equivalence condition which we must now discuss. The arguments occur most frequently in the context of the literature of discussion of the paradoxes of confirmation. We shall have occasion to treat more fully of these in Chapter X. However it will be useful to begin to discuss them here in order to discuss the arguments against the hypothesis equivalence condition.

Hempel's paradoxes of confirmation were first brought to our notice in an article in 1937 and then, more fully, in 1945 in [2]. The paradoxes are generated by the fact that three highly plausible principles of confirmation prove incompatible.

The first principle is known as 'Nicod's criterion'. It states that evidence of the existence of a ϕ which is ψ confirms 'all ϕ's are ψ', that is, with given background knowledge renders 'all ϕ's are ψ' more likely to be true than does background knowledge alone; evidence of the existence of a ϕ which is $\sim\psi$[1] disconfirms 'all ϕ's are ψ', that is, with given back-

---

(1) Henceforward I use '$\sim\psi$' to symbolize 'not-ψ', '$\sim\phi$' to symbolize 'not-ϕ' and so on.

ground knowledge renders 'all ϕ's are ψ' less likely to be true than does background knowledge alone; evidence of the existence of a $\sim\phi$ which is ψ, or a $\sim\phi$ which is $\sim\psi$ is irrelevant to, that is neither confirms nor disconfirms, 'all ϕ's are ψ'. Put symbolically '$\phi a . \psi a$' confirms, '$\phi a . \sim\psi a$' disconfirms, while '$\sim\phi a . \sim\psi a$' and '$\sim\phi a . \psi a$' are irrelevant. Nicod's criterion only concerns the effect of these four pieces of evidence on the hypothesis 'all ϕ's are ψ'. When putting forward this criterion ([3] p. 219) Nicod allowed that other pieces of information – e.g. evidence that a certain object very like a ϕ was not a ψ – could also confirm or disconfirm the hypothesis. Neither Nicod nor Hempel explicitly discuss under what conditions Nicod's criterion is supposed to hold, but, since their discussion does not suggest any limitation to its operation, we will suppose that it is meant to apply whatever the background evidence (unless, presumably, the background evidence entails 'all ϕ's are ψ' or its negation). Whether Nicod's criterion is correct is something which we shall have to investigate in Chapter X.

The second principle is usually referred to in the literature as 'the equivalence condition', but to avoid confusing it with the two axioms which are the main topic of this chapter, I will call it simply 'the c-condition'. This is that whenever evidence e confirms (that is, increases the probability of) a hypothesis h for some background evidence k, it also confirms h' for background evidence k, if '$h \equiv h'$' is analytically true; in symbols if '$h \equiv h'$' is analytically true and $P(h/k . e) > P(h/k)$, then $P(h'/k . e) > P(h'/k)$. The hypothesis equivalence condition entails the c-condition; and, although the converse does not hold, it is difficult to envisage someone maintaining the c-condition without maintaining the hypothesis equivalence condition.

The third principle is not given a name by Hempel, but I shall term it 'the scientific laws condition'. This states that a hypothesis 'all ϕ's are ψ' is logically equivalent to 'all $\sim\psi$'s are $\sim\phi$' and to 'everything which is ϕ or $\sim\phi$ is either ψ or $\sim\phi$'. Hempel professes to be considering what he terms 'general empirical hypotheses', that is what I have termed 'nomological propositions' and his 'all ϕ's are ψ' is therefore to be understood as 'of physical necessity all ϕ's are ψ', and the other cited propositions are to be understood in a similar way.

Different paradoxes are generated by considering different claims of the scientific laws condition. Thus given the c-condition and the first claim of the scientific laws condition, whatever confirms 'all ravens are black' also confirms 'all non-black objects are non-ravens'. But this contradicts Nicod's criterion, which claims that (if we use 'R' for raven and 'B'

for black) '*Ra . Ba*' confirms the former but is irrelevant to the latter, while '~*Ra . ~Ba*' confirms the latter but is irrelevant to the former. Given the *c*-condition and the second claim of the scientific laws condition, whatever confirms 'all ravens are black' also confirms 'everything which is a raven or no raven is either black or no raven'. But this contradicts Nicod's criterion, which claims that '~*Ra . Ba*' is irrelevant to the former but confirms the latter. These contradictions can be removed by dropping any one of the three principles, and hence solutions of the paradoxes can be classified according to which principle they propose to drop.

I will begin by considering the solution which involves dropping the scientific laws condition. The normal argument for this condition is that given by Hempel. According to Hempel, 'the meaning of general empirical hypotheses, such as that all ravens are black, or that all sodium salts burn yellow, can be adequately expressed by means of sentences of universal conditional form,[1] such as "(x) (Raven $(x) \supset$ Black (x))" and "(x) (Sod. Salt $(x) \supset$ Burn yellow (x))", etc.' ([2], pp. 15f). For clearly, since by the rules of logic '(x) $(Rx \supset Bx)$' is logically equivalent to '(x) $(~Bx \supset ~Rx)$' and to '(x) $((Rx \vee ~Rx) \supset (~Rx \vee Bx))$' the criterion follows. (I assume in this symbolization, and shall continue to assume, that the domain of x includes objects at all instants of time.)

However, Hempel's argument fails because '(x) $(\phi x \supset \psi x)$' does not adequately express the meaning of 'of physical necessity all ϕ's are ψ'. '(x) $(\phi x \supset \psi x)$' is a proposition which is true if, of every x, '$\phi x \supset \psi x$' is true; but is false of there is an x of which '$\phi x \supset \psi x$' is false. Now '$\phi x \supset \psi x$' will be true of every x, if no x's are ϕ; and so '(x) $(\phi x \supset \psi x)$' is automatically true if there are no ϕ's. However, as we noted in Chapter I, some nomological propositions 'of physical necessity all ϕ's are ψ' will be true and some will be false, when there are no ϕ's. The nomological proposition 'all ideal gases have $pv = RT^3$' is false, even though there are no ideal gases. No nomological proposition 'of physical necessity all ϕ's are ψ' is true just because there are no ϕ's; whereas '(x) $(\phi x \supset \psi x)$' is true if there are no ϕ's, just for that reason. Hence the one does not adequately express the meaning of the other, and Hempel's argument for the scientific laws condition fails.

(1) '( )' is known as the universal quantifier. It governs the following bracket and means 'for all x'. The domain of the quantifier is the set of objects which could be x. Thus '(x) (ϕx)' means 'for all x, x is ϕ', which is to say 'everything is ϕ' (everything within the domain of the quantifier, that is). '(x) $(\phi x \vee \psi x)$' means 'everything is ϕ or ψ'. '(x) $(\phi x \supset \psi x)$' means 'for all x, $\phi x \supset \psi x$', and so on.

Nevertheless, despite the failure of Hempel's argument, the scientific laws condition seems to me correct. 'All ravens are black' which is, let us remember, to be taken in the sense (1) 'of physical necessity all ravens are black', does seem to entail and to be entailed by (2) 'of physical necessity all non-black things are no ravens' for the following reason. Objects in the world can be divided into two categories – black and non-black. (1) claims that of physical necessity any raven, if one existed, would have to belong to the black category. This is to say that of physical necessity the non-black category would have to be empty of ravens, whether or not there were any ravens, which is to say (2). (2) expresses the 'general empirical hypothesis' or nomological proposition 'all non-black objects are non-ravens'. By a similar argument (1) is equivalent to (3) 'of physical necessity anything which is a raven or no raven is black or no raven' and hence to the 'general empirical hypothesis' or nomological proposition 'everything which is a raven or no raven is either black or no raven'. What I have discussed with relation to 'all ravens are black' clearly applies *mutatis mutandis* to all nomological propositions of the form 'all ϕ's are ψ'. Hence I conclude that the scientific laws condition is correct, although not for the reasons which Hempel gives.

With the scientific laws condition defended we are now in a position to consider the solution which consists in denying the c-condition. Since the hypothesis equivalence condition entails the c-condition, the falsity of the latter entails the falsity of the former. It is because attempts have been made to solve the paradoxes by a method which entails the denial of the hypothesis equivalence condition, that I have begun to treat of the paradoxes in this chapter. The proposed solution denies that it follows from 'e confirms h_1' and 'h_1 is logically equivalent to h_2' (i.e. '$h_1 \equiv h_2$' is a logical or analytic truth) that 'e confirms h_2'. To suggest this is not to suggest that in some way confirmation breaks the rules of logic. It is only to suggest that 'confirms' shares a property possessed by many intentional verbs such as 'hopes for', 'fears', or 'believes', that when the clauses following them in some proposition are replaced by logically equivalent clauses, a new proposition, not logically equivalent to the former, is obtained. 'John is 6 feet high' is logically equivalent to 'John is 72 inches high'. But 'George believes that John is 6 feet high' is not logically equivalent to 'George believes that John is 72 inches high'. Of the latter pair, one could be true when the other was false.

The suggestion that it is the c-condition which is at fault originates in some hints of Goodman [4] which have been developed by Morgenbesser [5] and Scheffler [6]. Scheffler's treatment is the fullest, and so I will consider it.

Scheffler claims that confirmation of a hypothesis is 'selective confirmation' against its contrary, that is, that you confirm a hypothesis, at any rate a hypothesis of the form 'all ϕ's are ψ' by producing evidence which is compatible with it but which is incompatible with its contrary. Scheffler has a narrow understanding of 'the contrary' of a hypothesis. In the wider and, I suspect, nowadays more usual understanding of the term, a proposition p is a contrary of a proposition q, if p and q cannot both be true together. A proposition will have many contraries; no one proposition is '*the* contrary' of it. In Scheffler's sense however, which has application only to propositions of the form 'all ϕ's are ψ', the contrary of 'all ϕ's are ψ' is 'all ϕ's are $\sim\psi$' (i.e. 'no ϕ's are ψ'). On this understanding of 'the contrary' the contrary of h_1 will not in general be logically equivalent to the contrary of h_2 if h_1 is logically equivalent to h_2. The contrary of 'all R's are B' is 'all R's are $\sim B$'. The contrary of 'all $\sim B$'s are $\sim R$' is 'all $\sim B$'s are R'. But 'all R's are $\sim B$' is not logically equivalent to 'all $\sim B$'s are R'. So on Scheffler's view what confirms 'all R's are B' is what is incompatible with 'all R's are $\sim B$', viz. '$Ra . Ba$', '$\sim Ra \sim Ba$' and '$\sim Ra . Ba$' are irrelevant since they are compatible with both 'all R's are B' and 'all R's are $\sim B$' (and, Scheffler assumes, equally probable on either hypothesis). Hence Nicod's criterion is preserved. However what confirms 'all $\sim B$'s are $\sim R$' is what is incompatible with 'all $\sim B$'s are R', viz. '$\sim Ba . \sim Ra$'. '$Ra . Ba$' and '$\sim Ra . Ba$' are irrelevant, since they are incompatible with neither 'all $\sim B$'s are $\sim R$' nor its contrary (and, Scheffler assumes, equally probable on either hypothesis). Hence the c-condition does not hold universally.

Now Scheffler's argument will only work if to confirm 'all ϕ's are ψ' is always[1] to rule out its contrary. For if sometimes one can confirm in other ways, then for all Scheffler has shown '$\sim Ra . \sim Ba$' could confirm 'all R's are B'. Now Scheffler may well be right that one adds to the likelihood of a hypothesis, that is confirms it, by ruling out rival hypotheses; or, to make his claim weaker, by reducing the probability of rival hypotheses. But it does not seem essential to the confirmation of a hypothesis that it be achieved by ruling out, or even reducing the probability of, its contrary. Hypotheses can be confirmed by evidence which confirms their contraries equally well. Evidence that each of many observed birds of many species had the same colour as other birds of its own species would confirm both 'all ravens are black' and 'all ravens are nonblack', by rendering less probable than before 'some ravens are black and

---

(1) Scheffler writes ([6] p. 291): 'The confirming of a hypothesis perhaps typically involves the favouring of it in this way against a contrary one.' Later passages show less hesitation.

some ravens are non-black'. Further, if confirmation did normally proceed by ruling out (as opposed to merely reducing the probability of) the contrary hypothesis, what would be the point of studying more than one raven in the process of assessing the confirmation of 'all ravens are black'? For the observation of one black raven suffices to rule out once and for all 'all ravens are non-black'. If we can go on to add to the confirmation of 'all ravens are black' by finding more black ravens, as is ordinarily supposed, it cannot be by ruling out more fully than we have done 'all ravens are non-black'. It must be by ruling out (or reducing the probability of) other rival hypotheses. I conclude that Scheffler is wrong to suppose that confirmation is always a matter of ruling out (or even reducing the probability of) the contrary hypothesis, and so that his argument against the *c*-condition, which depends on this supposition, is mistaken. Hence it cannot be used as an argument against the hypothesis equivalence condition which entails the *c*-condition. If we accept both the scientific laws condition and the *c*-condition, this means that we must solve the paradoxes of confirmation by rejecting Nicod's criterion. I believe that this is the correct solution of the paradoxes and I will argue for it in Chapter X.

However, I will now produce a different argument against both the hypothesis equivalence condition and the *c*-condition, although for reasons which I will give later I do not think that even if this argument is accepted, it provides the whole solution to the paradoxes. The argument is as follows. Any analytic proposition *h* is logically equivalent to any other analytic proposition h'. This is because any world of which *h* is true, h' will also be true, and this is because they will both be true of all worlds. Hence if the hypothesis equivalence condition is correct, the probability of *h* on any evidence will be the same as the probability of h'. But now let h' be a very simple tautology of the form '$p \vee \sim p$'. The probability of h' on any evidence at all, we are inclined to say, is maximal. Whatever we know about the world, that is, whatever our evidence, it is certain that h'. On any evidence *e* $P(h'/e) = 1$. Now let *h* also be an analytically true proposition, but one requiring a great deal of mathematical sophistication to show its truth, for example, the four square theorem, that any positive integer is a square or the sum of two, three or four squares; that is, for any positive integer *n*, $n = w^2 + x^2 + y^2 + z^2$ has a solution where *w*, *x*, *y* and *z* are either positive integers or zero. This is in fact a theorem provable from the axioms of arithmetic and so logically necessary. However, like the mathematician Bachet de Meziriac who first discovered it, we may be unable to prove it and so not be certain of its truth, since our background evidence k is evidence that does not include h. However we can apparently produce evidence in its support. Bachet verified that it held for numbers up to 325. Let e be 'every positive integer up to 325 is a square or the sum of two, three or four squares'. Now surely, we may feel, it is the case that in these circumstances h is more likely to be true given e than without it. And also, we may feel, even given e, h is not as likely to be true as h'. So $P(h'/e.k) > P(h/e.k)$ even though h and h', being both analytical propositions, are logically equivalent. If these intuitions are right, the hypothesis equivalence condition is false. In that case the c-condition is false too, because e confirms h, but not h', although h and h' are logically equivalent.

This point leads to a similar objection to the evidence equivalence condition:

AXIOM VI   For all propositions p, q, and r, if $p \equiv q$ is analytically
true, $P(r/p) = P(r/q)$

This condition, unlike the hypothesis equivalence condition, has been subjected to very little criticism in the literature. The vast majority of writers on confirmation theory have taken it for granted. Like the hypothesis equivalence condition it is, superficially, very plausible. If p renders r to some extent likely to be true, then the state of the universe described by p renders the state described by r likely to that extent. But the state of the universe described by p is the state described by q. Hence q renders r to that extent likely to be true. Yet by the above argument, if 'k', 'h' and 'e' are as above, then $P(h/e.k) > P(h/k)$. Yet since e is true of logical necessity (it can be proved deductively from the axioms of arithmetic) '$e.k$' is logically equivalent to 'k'. (The conjunction of any proposition with a logically necessary proposition yields a proposition logically equivalent to the first proposition.)

Countless similar examples can be produced of mathematical propositions apparently rendered probable by evidence which is less than conclusive, and which therefore form apparent counter-examples to the two equivalence conditions. ([7] is a study of such evidence for mathematical hypotheses, or conjectures, as they are often called, and provides many examples.)

Against this it may be urged that the mathematical examples are examples of a somewhat special kind, since the hypotheses whose probability is being assessed are either necessary truths or their denial. To reinforce the argument I shall therefore produce counter-examples to the evidence equivalence condition of a somewhat different kind. In these

E

examples, as in the mathematical examples, it appears that the addition of an analytic truth to evidence affects the probability of an hypothesis, but in these examples the hypothesis is an ordinary scientific one, not a mathematical one. But since the conjunction of an analytically true proposition with any other proposition yields a proposition logically equivalent to the latter, the evidence equivalence condition is violated. In these cases what is added to the evidence about a scientific hypothesis is the analytic truth that a certain new theory if true would predict many of the observations which were evidence for the old theory.

In 1542 the evidence in the form of reports of planetary positions (k) apparently rendered it probable that in its main ideas the Ptolemaic Theory of Astronomy (h) was correct. This was because the theory (h) allowed men to predict many of the reports of planetary positions (k_1) from a few initial ones (k_2), k consisted of k_1 and k_2. h and k_1 entailed k_2. In 1543 was published Copernicus's *De Revolutionibus* in which Copernicus advocated a different (and possibly simpler) theory of planetary motion to the Ptolemaic Theory. Copernicus cited no new observations in favour of his theory, but showed that the old observations were to be expected on a very different theory of planetary motion. *De Revolutionibus* expounded the analytic truth that Copernicus's Theory h' (a synthetic theory) together with the observational reports k_1 also entailed the observational reports k_2; and that h' was a comparatively simple theory. This analytic truth e was then added to the body of scientific knowledge, to the evidence available to every investigator of planetary motion. As a result of Copernicus's work, it would, I suggest, be held by many people that the Ptolemaic Theory was less probable on the available evidence, less likely to be true, than it had been in 1542. Of course the events of 1543 are no isolated case. Countless times in the history of science the mere invention of a theory, and the demonstration that it is a comparatively simple theory which together with certain particular propositions predicts certain other particular propositions, all of which is the discovery of analytic truths, has apparently affected the probability of existing theories. This is because it has shown that old observations were to be expected given a different theory from those previously used, which makes it less likely that one of those latter theories provides the true explanation of the observations, and so is true. The invention of General Relativity in 1916 apparently decreased the probability of Newtonian theory being true. Yet all that Einstein showed at that time is that certain known observations were to be expected given certain simple axioms, which was an analytic truth. So does not the addition of an analytic truth to evidence affect the probability of a hypothesis? The supposition that it does is inconsistent with the evidence equivalence condition.

If these counter-examples are admitted, the equivalence conditions must go. Yet a defendant of these conditions could deny the description given in expounding the counter-examples, of the relation of evidence to hypothesis. He might claim that in cases where analytic propositions added to evidence appear to make a difference to probability, they do not do so really; only (falsely) appear to do so. What in fact happens is that the discovery of those analytic propositions helps us rightly to assess the evidence and so rightly to assess the true value of the probability. Thus in the mathematical cases it would be claimed that the probability of a mathematical proposition which is in fact true because it is an analytic truth and so (given the understanding of mathematical terms embodied in their definitions) deducible from any evidence, is on any evidence 1. Any false mathematical proposition will be the denial of an analytic truth and so will have on any evidence a probability 0. If we are unable to work out whether a mathematical proposition is an analytic truth or the denial thereof, then we do know what the probability is. In such a circumstance we will have to guess, and certain analytic truths (e.g. that the hypothesis holds for all integers up to 325) make our guesses reasonable or unreasonable. But the probability is unaffected by knowledge of such analytic truths. Again in the scientific cases it would be claimed that there is a true probability of any hypothesis on some evidence, but men may work it out wrongly, and discovery of analytic truths may draw their attention to their false evaluation of the probability. The probability of the four-square theorem was in fact, on any evidence, the same as that of any simple tautology. Yet only when a proof of the theorem was produced could men be sure of that. The probability of Ptolemy's Theory was in fact less than the men of 1542 believed. The discovery of Copernicus's Theory brought this to their notice.

Now I cannot see that ordinary usage provides a clear answer about what to say about such examples. Some people would say that the mathematical discovery of its proof made the four-square theorem more probable than it had been; and others would say that the discovery merely showed the true value of the probability. Some people would say that the invention of Copernicus's Theory and discovery of its simplicity and predictive power made Ptolemy's Theory less probable; others would say that the discovery revealed how comparatively improbable Ptolemy's Theory always had been.

As a philosopher describing the criteria of epistemic probability used by ordinary people and scientists, I could simply leave it at that. But whether or not the equivalence conditions hold seems to make such a difference to the truth or falsity of so many propositions of epistemic probability – for instance the theorems which we can derive from the first four axioms of the calculus alone are very limited. Hence it is worthwhile investigating what arguments can be produced in favour of making our concept of epistemic probability a more precise one in one or other of the possible ways – either by insisting on the equivalence conditions or by allowing analytic truths to affect probability.

In favour of the latter move is the fact that if we adopt it we do not have to say that the ancients so persistently misjudged probabilities as we would have to say if we held to the equivalence conditions; and we may not feel it fair to accuse them of much misjudgement. For if such ancient scientific theories as Ptolemy's Theory of Astronomy are not *shown* less probable by discoveries of rival theories (and of the facts that these are simple and can explain old observational evidence), only *made* less probable, then it could be that on the evidence available to the ancients they were quite probable, and so the ancients did not misjudge their probabilities. But if discovery of new theories revealed the improbability of the ancient ones, the ancients misjudged the probabilities of the latter. (Of course one major reason for holding many ancient theories improbable is that new observational or synthetic evidence has turned up – e.g. Galileo's observations with his telescope counting against Ptolemy Theory. The fact that such new evidence makes old theories less probable does not mean that they were improbable on the old observational evidence. All that is at stake here is the effect of analytic discoveries on the probabilities of ancient theories.) In general too, if analytic evidence is allowed to affect probability, one will be less likely to be radically wrong in one's judgements of probability than one might otherwise be – e.g. about the probability of a mathematical conjecture.

In favour of sticking by the equivalence conditions are the considerations given earlier in the chapter in their favour. With the equivalence conditions too epistemic probability becomes a much more objective matter. One reason for this is the following. Given the equivalence conditions, the probability of h on e and of h' on e' will be the same if h and h' are logically equivalent, and so are e and e'. It is comparatively easy to ascertain whether the propositions expressed by two sentences are logically equivalent – one has only to show that in any world in which one expressed a true proposition the other would too. But if the equivalence conditions are abandoned, it is important for us to be able to make a further distinction among sentences which express logically equivalent propositions, between those which express the same identical proposition and those which express different propositions. This is because, as we saw in Chapter I, probability belongs to propositions rather than sentences. The probability of a proposition is the same, whichever sentence is used to express it. A sentence in French expresses the same proposition as a sentence in German which translates it. Also two sentences in English can express the same proposition. 'Caesar was killed by Brutus' seems to express the same proposition as 'Brutus killed Caesar'. Yet not all sentences which express logically equivalent propositions express the same proposition. Presumably 'Brutus killed Caesar' and 'Brutus killed Caesar and $\int x^2 dx = \frac{1}{3} x^3$' express different propositions – they say different things. If the equivalence conditions are dropped, the probability of h on e and the probability of h' on e' are the same if h and h' are the same proposition, and e and e' are the same proposition, but not necessarily if these are pairs of non-identical but logically equivalent propositions. Yet we have no clear criteria here for when two sentences which express logically equivalent propositions express the same proposition. There are some clear cases of when they do and when they do not, but there are a vast number of cases where it is not clear whether they do or do not. Does 'the British Prime Minister met President Pompidou in 1971' express the same proposition as 'Mr Heath met President Pompidou in 1971'? We do not need to face these questions calling for answers which are hard to provide if we stick by the equivalence conditions. In that case if sentences express logically equivalent propositions, it does not matter whether or not they express identical propositions.

Such are some of the arguments for and against sticking by the equivalence conditions. If we do abandon the conditions it is plausible to replace them by two weaker conditions which surely are necessary truths. These are:

AXIOM V*   For all propositions p, q and r, when r includes the proposition that $p \equiv q$ is analytically true, $P(p/r) = P(q/r)$.

AXIOM VI* For all propositions p, q, r and k, when k includes the proposition that $p \equiv q$ is analytically true, $P(r/p \,.\, k) = P(r/q \,.\, k)$.

These axioms say that propositions *known* to be equivalent (whether or not of logical necessity) can be substituted for each other in probability expressions, while these retain their values; they do however allow for the possibility that the addition to the old evidence of the evidence that the propositions in question are equivalent may make a difference to the probability value. The previous axioms stated that propositions logically equivalent to each other (whether or not known so to be) can be substituted for each other, while these retain their values. None of the arguments used earlier counts against the new axioms.

I have given the arguments for and against the equivalence conditions, and claimed that choice between keeping them and abandoning them in favour of the weaker axioms sketched above is essentially a matter for decision rather than analysis of existing concepts. Yet in view of the fact that so much work on Confirmation Theory has proceeded within a tradition which took the equivalence conditions for granted, and that, as I wish to provide an introduction to Confirmation Theory, I do not wish to throw aside its main tradition, I shall take the equivalence conditions as axioms. If anyone denies them he will have to qualify any conclusions which I draw subsequently with their aid in the following way. He will have to say that the conclusions only hold if the logical equivalences relied on (that '$h \equiv h'$' or '$e \equiv e'$' are logically true in the case in question) are *known* to hold, i.e. that they hold is part of the evidence.

It was pointed out earlier in the chapter that the hypothesis equivalence condition entails the *c*-condition. Hence if we retain the equivalence conditions, the solution of the paradoxes of confirmation can only be found in the abandonment of Nicod's criterion. However even if the equivalence conditions are abandoned in favour of the weaker Axioms V* and VI* above, this will not provide a general solution to the paradoxes. For a man's evidence *e* may well include the scientific laws condition, or some part of it, say that *h* 'of physical necessity all ϕ's are ψ', is logically equivalent to h', 'of physical necessity all $\sim\psi$'s are $\sim\phi$'. In such case, by Axiom V*, $P(h/e) = P(h'/e)$ and hence whatever confirms *h*, i.e. raises its probability, confirms h'. In such a case, it cannot be that '$\phi a . \psi a$' alone confirms 'all ϕ's are ψ', and '$\sim\phi a . \sim\psi a$' alone confirms 'all $\sim\psi$'s are $\sim\phi$'s'. Hence Nicod's criterion cannot be in general true. If the modest piece of logical knowledge embodied in the scientific laws condition is included among our evidence, the solution to the paradoxes of confirmation, which still arise in this situation, must lie in the falsity of Nicod's criterion. So whichever decision we take about the equivalence conditions, the falsity of Nicod's criterion follows from the weaker Axiom V* above, together with the scientific laws condition. Nicod's criterion will be examined in detail in Chapter X, and the limits to its application there considered.

Bibliography

For Carnap's statement of the two equivalence conditions, see:

[1] RUDOLF CARNAP *Logical Foundations of Probability,* 2nd edition, London, 1962, p. 285.

For Hempel's exposition of the paradoxes of confirmation, see:

[2] C. G. HEMPEL 'Studies in the Logic of Confirmation' *Mind,* 1945, **54,** 1–26 and 97–121; reprinted with an additional postscript in C. G. HEMPEL *Aspects of Scientific Explanation,* New York, 1965, pp. 1–51. My page references are to the latter version.

For Nicod's criterion, see:

[3] JEAN NICOD *Foundations of Geometry and Induction,* London, 1930, pp. 219 ff.

For the attempt to solve the paradoxes by rejecting the hypothesis equivalence condition, see:

[4] N. GOODMAN *Fact, Fiction, and Forecast,* 2nd edition, Indianapolis, 1965, pp. 69–72.

[5] S. MORGENBESSER 'Goodman on the Ravens' *Journal of Philosophy,* 1962, **59,** 493–5.

[6] ISRAEL SCHEFFLER *The Anatomy of Inquiry,* New York, 1963, pp. 258–95. (The earlier pages contain an outline of the paradoxes and discussion of some solutions involving a rejection of Nicod's criterion – for which see my Chapter X. Scheffler's pp. 286–95 contain his own solution of rejecting the *c*-condition.)

For examples of non-conclusive evidence apparently supporting mathematical conjectures, see:

[7] G. POLYA 'Induction and Analogy in Mathematics' in *Mathematics and Plausible Reasoning,* London, 1954, Vol. I.

CHAPTER V

The Disjunctive and Conjunctive Axioms

So much for Axioms V and VI of the probability calculus of propositions. I argued in Chapter III for the correctness of Axioms I and II. What of the remaining axioms - III and IV?

AXIOM III   For all propositions p, q and r, if $\smallsmile(p \cdot q \cdot r)$ is analytically true,

$P(p \vee q/r) = P(p/r) + P(q/r)$.

AXIOM IV   For all propositions p, q and r

$P(p \cdot q/r) = P(p/q \cdot r) \times P(q/r)$.

It is generally agreed that these axioms, as well as the first two axioms, govern all probabilities $P(p/r)$ in which the hypothesis p is a particular proposition or a conjunction of a finite number of particular propositions. This seems highly plausible when we consider examples. Let r be 'one and only one card is drawn from a pack', p be 'a heart is drawn', q be 'a diamond is drawn', $P(p/r) = \frac{1}{4}$, $P(q/r) = \frac{1}{4}$. Surely then we would say $P(p \vee q/r) = \frac{1}{2}$ which is the result given by Axiom III. Again, let r be 'a biased dice is thrown on just one occasion', p be 'a six is thrown', q be 'a two is thrown', $P(p/r) = \frac{1}{2}$, $P(q/r) = \frac{1}{6}$. Then surely we would say $P(p \vee q/r) = \frac{1}{2} + \frac{1}{6} = \frac{2}{3}$, again in conformity with Axiom III. Let r be 'two cards are drawn without replacement from a pack', q be 'a heart is drawn first time', p be 'a heart is drawn second time', $P(q/r) = \frac{1}{4}$, $P(p/q \cdot r) = \frac{12}{51}$. Then surely $P(p \cdot q/r)$, that is the probability that both cards drawn are hearts, $= P(q/r) \times P(p/q \cdot r) = \frac{1}{4} \times \frac{12}{51} = \frac{3}{51}$. This is what Axiom IV claims.

However, can a better proof be given of the axioms than the above demonstration of their application to particular examples, and further, do the axioms apply when the hypotheses are nomological propositions? Many have doubted whether they do apply when the hypotheses are nomological propositions. Two kinds of proof of the axioms have been proposed by philosophers and mathematicians. The first (accepting an axiom virtually the same as my Axiom II as intuitively obvious)[1] attempts to show Axioms III and IV to be consequences of more fundamental principles which are intuitively obvious. Such an attempt was made by R. T. Cox [1] and a similar attempt has been made recently by J. R. Lucas [2].

Cox proves Axioms III and IV to be consequences of two fundamental principles, assumptions about certain functions being differentiable, and an arbitrary simplification, all given his Axiom II. The two fundamental principles are (i) that $P(\sim q/r)$ is a unique function of $P(q/r)$, and (ii) that $P(q \cdot r/s)$ is a function of $P(r/s)$ and $P(q/r \cdot s)$. On the reasonable supposition that such functions can be differentiated (twice) with respect to each other, he proves that given his equivalent of my Axiom II and his two principles, $P(q/r \cdot s) = P(r/s) \times P(q/r \cdot s)$. This is our Axiom IV, the conjunction axiom. He then shows it to be a consequence of what is so far assumed that $\{P(\sim q/r)\}' = 1 - \{P(q/r)\}'$. He then adopts the simplification that $r = 1$, and so obtains the negation rule $P(\sim q/r) = 1 - P(q/r)$. From the conjunction axiom and the negation rule follows the disjunction axiom, our Axiom III. Lucas proceeds on similar lines to Cox. He shows that we can prove that $r = 1$, if we adopt any one of three principles. Expressed in our notation and using, by (i) above, $F(P(q/r))$ to

represent $P(\sim q/r)$ and $F'(P(q/r))$ to represent $\dfrac{d}{dP(q/r)} F(P(q/r))$, these are

(iii) $F'(P(q/r)) = F'(F(P(q/r)))$,

or   (iii)' $F'(P(q/r))$ is always negative,

or   (iii)'' $P(q/r) = F(P(q/r))$ if and only if $P(q/r) = \frac{1}{2}$.

Of these (iii)'' seems intuitively obvious and so we need not bother about (iii) and (iii)'. (i) and (ii) are likewise hard to deny. Following on the work of Cox, Lucas has produced an argument which bases Axioms III and IV of the probability calculus on some very sure foundations.

The other kind of proof of Axioms III and IV of the probability calculus or propositions is of a radically different character. It has been shown (e.g. in [4]) that if a man A made judgements of probability about a number of propositions p, q, r etc. which do not conform to the first four axioms of the calculus, and so in particular to Axioms III and IV, and if he was prepared to take bets in accord with these judgements, then someone could make a 'Dutch book' against him. This is to say that

----

(1) The actual assumption which Cox makes is that for all i and h $P(i/h \cdot i) = 1$.

if A was prepared to take bets in accord with judgements of probability which did not conform to the axioms, then someone could lay money with him in such a way that A could not possibly win anything but might lose something. By A taking bets in accord with judgements of probability I mean the following: if A judges that $P(q/e) = x$, he is prepared to accept bets from anyone at the odds of $1 - x : x$ or less on q, given the truth of e. For example if A judges that, given that Nijinsky runs in the St Leger (e), the probability that he will win (q) is $\frac{1}{3}$, then he will be prepared to accept bets of $\frac{2}{3} : \frac{1}{3}$ (2 : 1) or less (e.g. 3 : 2 or 5 : 4 or 1 : 1 or 1 : 2) on Nijinsky winning. This is to say that if B places a bet of £1 with A on Nijinsky winning, A is prepared to agree with B to pay him up to £2 if Nijinsky wins (given that he runs) while B pays A £1 if Nijinsky loses. If a 'Dutch book' can be made against a man, his judgements of probability are said to be incoherent. What has been shown is that to be coherent our judgements of probability must conform to the axioms of the probability calculus. One example illustrating this point is the following. Suppose there are only three horses, X, Y and Z in a race, of which one must win, and A judges that the probability of X winning is $\frac{1}{4}$, and of Y winning is $\frac{1}{4}$, and of Z winning is $\frac{1}{4}$. If he takes bets in accord with these judgements, he will be willing to take bets at odds of 3 : 1 (or less) on each horse. But then if B puts £1 on each horse at 3 : 1, whatever horse wins, B will gain and so A will lose (£3 − £1 − £1) = £1. Hence A's judgements of probability are incoherent. Further, they violate Axiom III of the probability calculus, since by it the sum of the probabilities of alternatives which are exclusive and exhaustive is 1, whereas on A's judgements it is $\frac{1}{4} + \frac{1}{4} + \frac{1}{4} = \frac{3}{4}$. For the general proof that if a man bets in accord with judgements of probability which do not accord with the axioms of the calculus, his judgements will be incoherent, see, e.g. Shimony [3].

But given all this, it may be asked why should our judgements of probability be coherent, unless we bet in accordance with those judgements in order to win? Shimony answers that in essence 'any decision whose success or failure depends on the actual truth or falsehood of the propositions which are the objects of belief is a bet. Thus the decision to cross a street is a bet whose success depends on the truth or falsehood of the proposition that no accident will befall me while crossing' ([3] p. 7). Now certainly Shimony is right that if I perform an action to secure some reward R, there will be some proposition q such that necessarily if q is true, I obtain R, but if q is false I do not. Hence my action may be regarded as a bet on the truth of q with a certain reward if I succeed and a certain loss if I fail, gain and loss to be measured not in monetary terms but in terms of the worth to the agent of success or failure. Hence a man who acts on judgements of probability which are inconsistent with the axioms will find himself doing actions which can bring no overall gain to him. A man's actions will fail to bring gain unless the probability judgements on which they are based conform to the axioms. If a man does actions which are bound to bring him no net gain by whatever standard he estimates worth (e.g. personal benefit or benefit to the community), he will be acting so as to frustrate his own ends, which is to act irrationally.

Now this argument shows that to act rationally a man must judge that probability propositions conform to the axioms; not that probability propositions in fact do conform to the axioms, which is not quite the same. Further, although all actions can be regarded as bets on propositions, not all propositions can be regarded as the subjects of possible bets. The betting argument will only work for those propositions q such that there is some reward R, such that necessarily if q is true I obtain R and if q is false I do not. For many propositions q such a reward R is conceivable. However, if q is a nomological proposition there can be no reward R which necessarily I receive if q is true and do not receive if q is false. For any receivable reward I care to specify which I will receive if q is true I may also receive if q is false. For a nomological proposition to be false it is sufficient that its predictions should fail to come off at some time in the indefinite future or would have failed to come off under some circumstances which in fact failed to be realized. Hence any result you care to name R such that it is obtained if q is true may also be obtained if q is false. Suppose q is 'of physical necessity all A's are B'. A man promises to give you R, say £1, if the next A is B. Then indeed you receive R if q is true; but you may also receive it if q is false. The same applies to any other receivable reward which can be named. Consequently no action can be regarded as a bet on a nomological proposition. Hence there seems to be no argument why our judgements of probability about nomological propositions should be coherent. I conclude that the betting argument does not show that we must judge all propositions of epistemic probability to conform to the axioms of the calculus, only that we must judge many such propositions to conform. For these two reasons I conclude that the more straightforward proofs given by Cox and Lucas have far greater force in showing the truth of Axioms III and IV than does the betting argument.

There are however two difficulties about Axiom III which must now be tackled. Axiom III states that:

For all propositions p, q, and r, if $\sim(p . q . r)$ is analytically true, $P(p \vee q/r) = P(p/r) + P(q/r)$.

It follows from this for any finite set of n hypotheses h_i mutually incompatible on the evidence e, however large n is, that for h', the proposition that one of the hypotheses h_i is true, $P(h'/e) = \sum\limits_{i=1}^{i=n} P(h_i/e)$. (In words, the probability that one of the hypotheses h_i is true is equal to the sum of the probabilities of each being true.)

Now consider the case of an infinite number of hypotheses h_i mutually incompatible on evidence e, each of which is as probable on that evidence as each other one. One example of such a set might be the set of propositions about the height of the table in the next room on evidence that it is between 2 feet and 3 feet tall and that matter is infinitely divisible. Each hypothesis about its height attributing to it a height of x feet where x is one of the infinite number of real numbers between 2 and 3 would seem to be as probable as any other such hypothesis on the cited evidence. Now what is the probability on the cited evidence of any one such hypothesis, $h_1 - P(h_1/e)$? Either $P(h_1/e) = 0$ or $P(h_1/e) = r$ where r is some 'infinitesimal value greater than 0' or $P(h_1/e) = q$ where q is a finite value greater than 0. But in the latter case there will be a finite integer m such that $mq > 1$ but $(m-1)q \leqslant 1$. Then consider any m of the hypotheses h_i. By Axiom III the probability that at least one of these hypotheses is true will be m times the probability that any given one is true, viz. mq, which is greater than 1. But by Theorem 3 of the calculus the maximum value of a probability is 1. Theorem 3 is a consequence (see p. 41) of Axioms I, II and III of the calculus. So in view of the intuitive obviousness of Axioms I and II, Axiom III seems to lead to self-contradiction, if we adopt the supposition that $P(h_1/e) = q$, where q is a finite value greater than 0. However, this supposition seems clearly false. For we surely want to say that the probability of one of an infinite number of equiprobable hypotheses mutually incompatible on the evidence, one of which must be true, is less than that of one of any finite number of equiprobable hypotheses mutually incompatible on the evidence, one of which must be true. Now let m and q be as above. Then suppose that there are m equiprobable hypotheses H_i mutually incompatible on evidence, one of which must be true. It does seem right to apply Axiom III to this case and to conclude that the probability of each is $1/m$. But $1/m$ is less than q (since $mq > 1$). If we suppose that each of an infinite number of equiprobable hypotheses mutually incompatible on the evidence has some finite probability q, then we have to say that that probability is greater than that of each of some finite number of equiprobable hypotheses mutually incompatible on the evidence, one of which must be true. This seems clearly false.

The alternatives are to say that the probability of each of an infinite number of equiprobable hypotheses h_i mutually incompatible on the evidence is 0, or that it has some infinitesimal value r greater than 0. To say the latter would be to say that for any finite number m, $0 < mr < 1$. But to assert the existence of such numbers would be to introduce a new mathematics, for our ordinary mathematics does not countenance the existence of infinitesimal numbers of this character. The other possibility is to say that every hypothesis of the above type has a probability 0, even though the evidence is not incompatible with it. This is to say that on evidence e every such hypothesis h_i is as probable as a hypothesis incompatible with e or even as a self-contradictory hypothesis. At first glance this seems false, but is it? How probable is it that the height of the table in the next room is, among the infinite number of possible values of the height, exactly 2·5 feet, on evidence that it is between 2 feet and 3 feet in height and that matter is infinitely divisible? There is of course some positive probability that the height is very close to 2·5 feet, say within 0·00001 feet of 2·5 feet; but is there a positive probability that it is exactly 2·5 feet? This seems doubtful, especially in view of the fact that to whatever degree of accuracy one measured the height of the table there would always remain an infinite number of values within which the height could lie.

In reflecting on these two possible views about the probability of each of an infinite number of equiprobable hypotheses h_i mutually incompatible on the evidence, I am inclined to hold that the view that the probability has an infinitesimal value is the one which is most consonant with our ordinary thinking about probability. However in view of the impossibility of producing a new mathematics in the course of a work devoted to another topic, I shall somewhat hesitantly take the alternative that the probability of each of the hypotheses h_i is 0, and explore its consequences. It does not seem a *great* distortion of our ordinary views about probability to adopt this supposition in order to bring coherence into those views.

The adoption of the alternative that each of the above hypotheses h_i has probability 0 leads however to a further difficulty. It follows from Axiom III, as I pointed out earlier, that for any finite set of n hypotheses h_i mutually incompatible on evidence e, however large n is, that for h', the proposition that one of the hypotheses h_i is true, $P(h'/e) = \sum_{i=1}^{i=n} P(h_i/e)$.

It seems natural to suppose that this also holds where the number of
mutually incompatible hypotheses is infinite. We do however need a
separate axiom to state this, because it does not follow from Axiom III
in accordance with the normal strict rules of mathematical inference.
Hence it is natural to add to our axiom set the following:

AXIOM III*   Where h' is the proposition that one of the infinite
number of hypotheses h_i mutually incompatible on the
evidence e is true,

$$P(h'/e) = \sum_{i=1}^{i=\infty} P(h_i/e).$$

But in that case consider the infinite set of equiprobable hypotheses
mutually incompatible on the evidence, the probability of each of which,
we supposed, is 0. By Axiom III* the probability that one of these hypotheses is true will also be 0. But if the hypotheses are, as in the example
considered earlier, not merely exclusive but exhaustive (i.e. one of them
must be true, for example one of the hypotheses ascribing to the table
some particular height between 2 feet and 3 feet must be true) then the
probability that one of them is true is, by Theorem 1, 1. Theorem 1
follows directly from Axioms II and VI which we have already seen good
reason for adopting; and the result is anyway surely intuitively obvious.
Consequently Axiom III* must not be adopted if we suppose – as we
have done – that each of an infinite set of equiprobable hypotheses
mutually incompatible on the evidence has probability 0. If we managed
to give to each of such hypotheses an infinitesimal probability, we could
perhaps adopt Axiom III*. (In his betting argument for the axioms of the
probability calculus Shimony – [3] pp. 18ff – showed the incoherence
which would result from the adoption of Axiom III*. Kolmogorov
added it to his original axiom set, but it was not one of Koopman's
axioms nor derivable from his axioms. See Bibliography to Chapter III
for the works of these writers.)
     None of what has been said means that we have necessarily to ascribe
to each of an infinite set of hypotheses mutually incompatible on the
evidence, e.g. to each of a set of nomological propositions consistent
with finite observational data, a probability of 0. On the contrary we
can ascribe to *each* of such hypotheses a finite probability greater than
0. But we can only do this if no more than a finite number of such
hypotheses have the same probability as each other. We can for example
ascribe a finite probability to each of an infinite number of hypotheses

if they can be ordered in a series in some way, and probabilities apportioned to each in turn from members of a convergent series of numbers which sum to 1 or less. Thus if we number the infinite series of hypotheses by the numbers of the series of natural numbers $h_1, h_2 \ldots h_n \ldots$ and h_1 has a probability of $\frac{1}{2}$, h_2 of $\frac{1}{4}$, and generally h_n has a probability $\frac{1}{2^n}$, then each of the infinite number of hypotheses will have a finite probability, the sum of which probabilities is 1. We will always have, as we saw earlier with the curve fitting problem (p. 45), an infinite number of mutually incompatible hypotheses in the form of nomological propositions which are equally accurate in their predictions of the evidence. We want to say, as we shall see in more detail in Chapter VII, that the evidence gives considerably greater probability to the simpler ones. This is quite compatible with the axioms of the calculus, if we suppose that in some vague way nomological propositions can to some extent be ordered in respect of simplicity. Chapter VII will examine the problem of ordering in more detail. What we cannot do, when maintaining the axioms of the calculus, is to ascribe to each of an infinite number of hypotheses a probability greater than some particular finite value p greater than 0. For in that case for some finite integer n greater than 0, the probability, by Axiom III, of one of n of those hypotheses being true, will be greater than 1 (since however small p is, so long as it has a finite value greater than 0, for some n, $np > 1$).

There is a number of kinds of sets of infinite numbers of nomological propositions to each of which we must attribute zero intrinsic probability. In the first place we must attribute zero intrinsic probability to any hypotheses of the form '$Pr(Q/A) = p$', with the possible exception of '$Pr(Q/A) = 1$' and '$Pr(Q/A) = 0$' (and by similar arguments to any hypotheses of the form '$\Pi(Q/A) = p$' with the possible exception of '$\Pi(Q/A) = 1$' and '$\Pi(Q/A) = 0$'). As we saw in Chapter II p can take any value among the infinite number of values in the real number continuum between 0 and 1. That we must attribute to each such hypothesis zero intrinsic probability can be seen with the aid of Principle A (see p. 43). In it let e be our total evidence and k be a tautology. Then we can imagine an infinite number of situations in which any one such hypothesis h_1 is as accurate in its predictions as a different such hypothesis h_n. Let us suppose that our evidence e is that n A's have been observed (in the process of looking for A's – this to avoid complications to be discussed in Chapter X) and nq of them found to be Q, together with other knowledge of the world which in a non-technical sense is not relevant to whether or not A's are Q (and so does not make any hypo-

thesis of the cited kind more probable than any other). Then it follows
from the results on pp. 28ff that in general for different n and nq there
will be different hypotheses h_n which are equally accurate with any
given hypothesis. Consider the hypothesis h_1 '$Pr(Q/A) = \frac{2}{3}$'. Suppose two
A's are observed and one found to be Q. Then this result is as probable
given '$Pr(Q/A) = \frac{2}{3}$' as given '$Pr(Q/A) = \frac{1}{3}$'. (On each hypothesis the probability that one out of two observed A's will be Q is $\frac{4}{9}$, and *ex hypothesi*
other evidence is equally likely on either hypothesis.) Suppose now that
three A's are observed and one found to be Q. This result is as probable
given '$Pr(Q/A) = \frac{2}{3}$' as given '$Pr(Q/A) = p$' where $p \simeq 0\cdot09$, and so on.
There is no end to the number of situations for which we can imagine
h_1 to be equally accurate with a different hypothesis. For each such
situation we would say that h_1 is as likely to be true as its rival. But by
Principle A if $P(e/h_1 . k) = P(e/h_n . k)$ (the two hypotheses are equally
accurate) and $P(h_1/e . k) = P(h_n/e . k)$ (the two hypotheses are equally
probable on the total evidence), then neither $P(h_1/k) > P(h_n/k)$ nor
$P(h_n/k) > P(h_1/k)$ and hence $P(h_1/k) = P(h_n/k)$, k being a tautology, this
means that h_1 and h_n have equal intrinsic probability. But by the above
argument there is an infinite number of hypotheses h_n of which this is
true. Hence, having equal intrinsic probability with each other, they
must each have a probability of zero.

In this case it follows from Bayes's Theorem (see p. 42) that no observational evidence e can give any finite degree of probability to any
such hypothesis. It might appear however that if we study a number of
A's and find that the proportion of them which are Q is p_1 ($\Phi(Q/A) = p_1$),
then hypotheses '$Pr(Q/A) = p$' are more probable on this evidence as
$p \to p_1$. For example, if 1,000 thirty-year-old university teachers have
been studied and it was found that 400 of them died before they were
seventy while 600 survived, it might seem that on this evidence the
(complex statistical) probability of a thirty-year-old university teacher
dying before seventy was more likely to be 0·4 than to have any other
value, more likely to be 0·39 than to be 0·38, to be 0·38 than to be 0·37
and so on. But in that case consider an arbitrary value p_x of p lying between 0 and p_1. If h_x, the hypothesis that $Pr(Q/A) = p_x$ has a probability
greater than the hypothesis that $Pr(Q/A) = 0$ and so a probability greater
than 0, and for each h_i, $Pr(Q/A) = p_i$, for $p_1 > p_i > p_x$, the probability
that that hypothesis is true is greater than the probability that h_x is true,
then we have an infinite number of hypotheses having a probability
greater than some finite value greater than 0. This, as we saw, is a supposition incompatible with the axioms of the calculus.

However, is it really more probable in such a case that $Pr(Q/A)$ is exactly p_k than that it is exactly p_m where p_k and p_m are values lying between p_x and p_1 above, p_k being nearer to p_1? Is it really more probable that the (complex statistical) probability of a thirty-year-old university teacher dying before seventy is *exactly* 0·38 than that it is *exactly* 0·37 (among the infinite number of possible values) on the above evidence that 400 out of 1,000 such university teachers have been found to die before the age of seventy? Certainly the true value of this (complex statistical) probability is more likely to be very close to (say within 0·0001 of) 0·38 than to be very close to 0·37, but that is an entirely different claim, and one quite compatible with the axioms of the calculus. However, if we did adopt the supposition that the true value was more likely to be p_k than p_m where p_k was a value lying closer to the observed proportion than p_m, then for the reasons given earlier we should have either to introduce infinitesimal probabilities or to abandon some axiom of the calculus, presumably Axiom III.

Principle C which stated that e confirms h if and only if $P(e/h.k) > P(e/k)$, and Principle B which stated that of two hypotheses with equal prior probability the more accurate one h was rendered more probable by the evidence, were both subject to the restriction 'given $P(h/k) \neq 0$'. The operation of the restriction is exemplified above.

We may note that the hypotheses to which we have been compelled, possibly with some reluctance, to attribute probability 0 on the evidence are peculiar in that, as we have interpreted them, no additional evidence could possibly show them to be true conclusively or beyond reasonable doubt. Nothing can show conclusively that $Pr(Q/A)$ is 0·05 rather than some value extremely close to 0·05. Hence nothing which has been said in the last few pages casts doubt on the claim made in Chapter III that the probability of a proposition which could be shown to be true is not 0.

The arguments of the last few pages have been directed to showing that we must attribute zero intrinsic probability to all hypotheses '$Pr(Q/A) = p$' (with the possible exception of '$Pr(Q/A) = 1$' and '$Pr(Q/A) = 0$') and to all hypotheses '$\Pi(Q/A) = p$' (with the possible exception of '$\Pi(Q/A) = 1$' and '$\Pi(Q/A) = 0$'). These arguments do not cover the possible exceptions. Thus '$Pr(Q/A) = 1$' is *never* in the simple situations discussed on p. 74 equally accurate with any other hypothesis '$Pr(Q/A) = p$', unless it is incompatible with the evidence e and in that case Principle A does not apply. For if n A's are examined, only if all n of them are found to be Q is '$Pr(Q/A) = 1$' compatible with the

F

evidence, and in that case it is more accurate than any other hypothesis '$Pr(Q/A) = p$'. Similar arguments apply to '$Pr(Q/A) = 0$', '$\Pi(Q/A) = 1$', and '$\Pi(Q/A) = 0$'. (Henceforward I will largely ignore the cases of '$Pr(Q/A) = 0$' and '$\Pi(Q/A) = 0$', since '$Pr(Q/A) = 0$' says the same as '$Pr(\sim Q/A) = 1$' and '$\Pi(Q/A) = 0$' says the same as '$\Pi(\sim Q/A) = 1$'. Results for the latter kind of case can therefore be applied to the former.)

'$\Pi(Q/A) = 1$' says 'of physical necessity all A's are Q'. It is natural to suppose that may such universal nomological propositions (and so propositions '$Pr(Q/A) = 1$', since '$Pr(Q/A) = 1$' is entailed by '$\Pi(Q/A) = 1$') have intrinsic probability greater than 0. This is because only if universal nomological propositions such as '$\Pi(Q/A) = 1$' have intrinsic probabilities greater than 0 can they have their probabilities raised, that is can they be confirmed, by observational evidence. (Since, as we saw on pp. 48 for such evidence $P(e/k) \neq 0$, it follows from Bayes's Theorem (p. 42) that if $P(h/k) = 0, P(h/e \, . \, k) = 0$.) But we do in general suppose that the universal nomological propositions which scientists put forward such as that all material bodies always have velocities less than or equal to c, the velocity of light, or that all water is composed of hydrogen and oxygen are confirmed by their predictions succeeding. Hence, given the axioms of the calculus, these universal nomological propositions cannot have intrinsic probabilities of 0.

The supposition that '$\Pi(Q/A) = 1$' and '$\Pi(Q/A) = 0$' have greater intrinsic probabilities than all other propositions of the form '$\Pi(Q/A) = p$' is made plausible by an important difference which holds between them if 'Q' is an observable property. By 'Q' being an observable property I mean that it is such that we can (at any rate sometimes) come to know whether an object is Q. Having a velocity less than or equal to c, being green, being composed of hydrogen and oxygen, being a swan, weighing more than two tons, are observable properties of bodies. Given then that 'Q' is an observable property, it is the case that although observational evidence cannot conclusively show any proposition of the form '$\Pi(Q/A) = p$' to be true, it can show two and only two such propositions conclusively to be false – viz. '$\Pi(Q/A) = 1$' and '$\Pi(Q/A) = 0$'. If we observe an A which is Q, that conclusively falsifies the second, and if we observe an A which is $\sim Q$ that conclusively falsifies the first. The same applies to propositions of the form '$Pr(Q/A) = p$'.

These various considerations lead me to conclude that (with exceptions which we must bring out shortly) propositions of the form '$\Pi(Q/A) = 1$' and '$\Pi(Q/A) = 0$' have probabilities greater than 0 while all other propositions of the form '$\Pi(Q/A) = p$' have probabilities of 0;

and that propositions of the form '$Pr(Q/A) = 1$' and '$Pr(Q/A) = 0$' have probabilities greater than 0 while all other propositions of the form '$Pr(Q/A) = p$' have probabilities of 0.

However several confirmation theorists have supposed otherwise. Both Carnap (see [5]) and Popper (see e.g. *The Logic of Scientific Discovery,* London, 1959, p. 363) claim that, given as they suppose that 'probabilities' conform to the axioms of the calculus, the intrinsic 'probability' of *every* nomological proposition and so its posterior 'probability' on evidence which consists of a finite number of particular propositions, are both 0. An advantage of Hintikka's system [6] is that such probabilities are not necessarily or in general 0.

We can neglect Popper's view since, as we shall see when we come to discuss Popper's account of acceptability in Chapter XIII (see p. 200) he understands 'probability' not in the ordinary sense, but in his special sense of 'logical probability'. In Carnap's system [5] in which 'probability' is understood in the ordinary sense the cited result follows from the adoption of his preferred function for assessing the probability of hypotheses, c^*. Carnap considers but dismisses alternative functions as less satisfactory. The characteristic of c^* is that it assigns equal intrinsic probability to each structure description of the world. A structure description is one which states the proportion of individuals in the world which have each combination of the possible primitive properties. (The primitive properties, according to Carnap, are 'qualitative' properties, and he assumes that these are recognized as such in advance before his system can be applied.) Thus if there are two primitive properties 'Q' and 'R', the structure description of the world states how many individuals are both Q and R, how many are both Q and $\sim R$, how many are both $\sim Q$ and R, and how many are both $\sim Q$ and $\sim R$. Hence given a world in which there is an infinite number of individuals, there will be an infinite number of structure descriptions which could apply to the world. Even if there is only one primitive property P, which an individual may either have or lack, if there is an infinite number of individuals, there will be an infinite number of possible ratios of P's to $\sim P$'s, and so an infinite number of structure descriptions which could apply to the world. If each such structure description has the same intrinsic probability as each other one, then for the reasons which I have given the intrinsic probability of each will be 0. Given that there is an infinite number of instants of time our world can be treated as one in which there is an infinite number of individuals, by treating objects at different instants of time as distinct individuals. (This will be appropriate since at different instants of time an object may have different properties, and in Carnap's system an individual always has the same properties.) Hence on Carnap's preferred system every 'hypothesis of universal form' about our world will have an intrinsic probability of 0. However Carnap's advocacy of c^* seems mistaken for a reason similar to that given earlier for '$\Pi(Q/A) = 1$' (and '$\Pi(Q/A) = 0$') often having greater intrinsic probability than other hypotheses of the form '$\Pi(Q/A) = p$'. A structure description in which many combinations are empty (e.g. silver which does not dissolve in nitric acid, material bodies which are attracted by other material bodies but do not themselves attract, copper which does not conduct electricity) and others full, seems to be a simpler description of the world and so to have greater intrinsic probability than other descriptions. Intuitively, evidence equally to be expected on the simpler as on more complex hypotheses makes the former more likely to be true. Scientists expect to find some properties invariably correlated with certain others and some properties never correlated with certain others.

Nevertheless, despite the intuitive plausibility of the claim that universal nomological propositions are often confirmed and so have greater than zero intrinsic probability, some writers have claimed that our intuitions are here misguided. They claim that we will see if we reflect on the matter that universal nomological propositions are not really confirmed by instances, since they purport to apply at all instants of an infinite time and at all points of a possibly infinite space. How, they argue, can such a vast claim be supported by observations in a tiny finite spatio-temporal period? What is confirmed is merely that some universal nomological proposition applies in a limited spatio-temporal region (as Hesse has argued in [7]) or that what it predicts on a given occasion is correct (as Carnap has argued in [5] pp. 571–4). However our *normal* intuitions, I suggest, do support the claim that many nomological propositions have greater than zero intrinsic probability.

Where the intrinsic probability of '$\Pi(Q/A) = 1$' is greater than that of other propositions of the form '$\Pi(Q/A) = p$' (apart from '$\Pi(Q/A) = 0$'), I shall say that '$\Pi(Q/A) = 1$' ('of physical necessity all A's are Q') is projectible (and similarly where 'Pr' is substituted for 'Π'). The term 'projectible' is due to Nelson Goodman, whose account of which nomological propositions are confirmed by observations will be examined in Chapter VII. For Goodman a projectible hypothesis is one confirmed by the success of its predictions of observations (for given background knowledge). If 'all A's are Q' (h) is confirmed by looking for and finding an A and observing that it is Q, then 'all A's are Q' is for Goodman projectible. As we saw on pp. 48f. 'all A's are Q' (h) will be confirmed by finding of a known A that it is Q if and only if $P(h/k) \neq 0$. Hence we can call a hypothesis projectible in Goodman's sense if its intrinsic probability is not zero.

Now we cannot suppose that all universal nomological propositions are projectible. One kind of proposition 'of physical necessity all A's are Q' which seems not to be projectible is one where the 'Q' records an exact numerical value of a measurement of a quantity which can vary continuously.

One nomological proposition where the consequent predicate 'Q' records an exact numerical value of a measurement of a quantity which can vary continuously is

h_1: of physical necessity all light travels at a velocity of exactly 300,000 km/sec.

h_1 is only one of an infinite number of hypotheses of the form

h: of physical necessity all light travels at a velocity of exactly j km/sec (where j is one particular real number between 299,998 and 300,002).

Suppose now that we have evidence that on a certain occasion light from a certain source travelled with a velocity of 300,000 ± 2 km/sec. (The exact velocity was not observed; what was observed was that it lay within the range stated.) Intuitively for each such hypothesis h $P(h/e . k)$, k being a tautology, is the same. On the stated evidence it is no more likely that the velocity of light is 300,000·01 km/sec than that it is 300,000·0132 km/sec. Also for any such hypothesis $P(e/h . k)$ is the same. $P(e/k) \neq 0$ and $P(e/h . k) \neq 0$ since e is evidence of observation compatible with h and k. Hence by Principle A, $P(h/k)$, that is $P(h)$, is the same for each of the infinite number of hypotheses of the stated form, and so is zero. That each such hypothesis has zero intrinsic probability is supported by the fact that 'travels at a velocity of exactly 300,000 km/sec' is not an observational predicate. No evidence of observation could show that the velocity of a particular shaft of light was exactly 300,000 km/sec rather than any other value extremely close to it. Hence h_1, and other universal nomological hypotheses attributing to all objects of some kind an exact numerical value of a measurement of a quantity which can vary continuously, differ in an important respect from many other universal nomological propositions. What apparently does not have zero intrinsic probability, because it can be confirmed by observations is

h': of physical necessity all light travels at a velocity 300,000 ± 2 km/sec.

Further, I suggest, though in this case somewhat hesitantly, universal nomological propositions 'of physical necessity all A's are Q' are not projectible if 'Q' or 'A' is a positional predicate. By a predicate 'Q' being positional is meant, roughly, that to say that an object is Q is to tell of its relations to some particular instant of time, point of space, physical object, or other particular thing. Positional predicates are often contrasted with qualitative ones which attribute to objects properties other than relations to particular instants of time, points of space, physical or other objects. Intuitively, we would say that 'found in Japan' or 'Victorian' are positional predicates, whereas 'made of oak' or 'very old' are not. A more rigorous definition of positionality will be provided in Chapter VII. The most famous example of a positional predicate discussed by philosophers is Goodman's predicate 'grue' (for bibliographical references, see Chapter VII) which I shall define as follows:

(x) {x is grue $\equiv$ ((x is green. x is at a time $<$ AD 2000) v (x is blue. x is at a time $\geq$ AD 2000))}

It will be seen that on this definition any object before AD 2000 which is green is also grue. Emeralds, grass, and ivy leaves today are grue. Any object after AD 2000 which is blue will be grue. If sea and sky remain blue after AD 2000 they will be grue. The concept of 'grue' and the literature about it will be examined in detail in Chapter VII.

An argument why we must suppose that any universal nomological proposition such as '(of physical necessity) all emeralds are grue' having a positional predicate as antecedent or consequent has zero intrinsic probability is that we do not regard such a hypothesis as confirmed when its predictions in the form of observational reports (deduced from the nomological proposition and background evidence) come off. From this it follows by the arguments of pp. 48f. that the nomological proposition has zero intrinsic probability. 'All emeralds are grue' predicts of an emerald observed in 1972 that it is green. If the prediction comes off, we do not regard 'all emeralds are grue' as any more likely to be true than it was. And, if we reflect on the matter, I suggest that we will see that the same still holds even if an emerald in AD 2001 is found to be blue. For in that case we would feel that there must be some reason why it and perhaps other emeralds are now blue instead of green other than a mere change of date – e.g. because of a change in our eyes, in the chemical make-up of emeralds, or something of that kind. Science does not postulate as laws propositions which refer to the particular; it postulates necessary connexions between things in virtue of their non-positional qualities. If all the rocks of a certain type are found in Japan, that has no tendency to show that there is any physically necessary connexion between being of that type and being found in Japan, only between being of that type and being found in an environment similar in some respect to that of Japan.

This argument is, I realize, less than conclusive, and I can see that in certain circumstances (e.g. in the year AD 2000 if emeralds suddenly turned blue for no other reason apparently, than that the year had changed) some people might judge a universal nomological hypothesis containing a positional predicate to be confirmed by observations. However, many such hypotheses (e.g. 'all emeralds are grue') clearly have less intrinsic probability than many other universal nomological propositions (e.g. 'all emeralds are green') even if the former probability is not zero, and rules from which this consequence results will be considered in Chapter VII. Chapter VII will also consider an attack by Goodman on the claim that universal nomological propositions containing positional predicates are not projectible, on the ground that there is no rigorous distinction to be made between positional and qualitative predicates.

This chapter has been concerned to expound and justify the disjunctive and conjunctive axioms of the probability calculus. We saw that we could only maintain the disjunctive axiom if we supposed that various propositions compatible with their evidence had on it a probability of zero. This led to an examination of which nomological propositions had an intrinsic probability of zero. We saw that our ordinary understanding of epistemic probability did not yield completely clear results here; but in so far as it did yield clear results, we have attempted to set them down. We concluded that universal nomological propositions in general have an intrinsic probability greater than zero, unless they are of one or two types which I have described. It may be that there are types of universal nomological proposition, other than those which I have described, which have zero intrinsic probability. Nevertheless I will henceforward assume that apart from propositions of the two types which I have described, all universal nomological propositions have intrinsic probability greater than zero.

Bibliography

For attempted proofs of the axioms of the probability calculus from more fundamental principles, see:

[1] RICHARD T. COX *The Algebra of Probable Inference*, Baltimore, 1961, Chapter I 'Probability'.

[2]   J. R. LUCAS *The Concept of Probability,* Oxford, 1970, Chapters 3 and 4.

For argument to the axioms of the probability calculus from coherence, see (e.g.)

[3]   ABNER SHIMONY 'Coherence and the Axioms of Confirmation' *Journal of Symbolic Logic,* 1955, **20**, 1–28.

The coherence argument was originally put forward in the context of the Subjective Theory of probability by F. P. Ramsey and B. de Finetti. See:

[4]   F. P. RAMSEY 'Truth and Probability' (first published 1931) and B. DE FINETTI 'Foresight: its logical laws, its subjective sources' (first published 1937), reprinted in H. E. Kyburg and H. E. Smokler (eds.) *Studies in Subjective Probability,* New York, 1964.

For Carnap's function c^* and his claim that the probability of 'hypotheses of universal form' is always zero, see:

[5]   RUDOLF CARNAP *The Logical Foundations of Probability,* 2nd edition, London, 1962, Appendix.

For Hintikka's system, see:

[6]   JAAKO HINTIKKA 'A Two-Dimensional Continuum of Inductive Methods' in Jaako Hintikka and Patrick Suppes (eds.) *Aspects of Inductive Logic,* Amsterdam, 1966.

For a discussion of whether nomological propositions have a probability (or as she calls it 'confirmation') of zero, see:

[7]   MARY HESSE 'Confirmation of Laws' in S. Morgenbesser, P. Suppes and M. White (eds.) *Philosophy, Science and Method,* New York, 1969.

CHAPTER VI

Further Objections to the Probability Axioms

The last three chapters have been concerned to interpret and to justify the claim that propositions of epistemic probability conform to the axioms of the probability calculus. This view is sometimes (not very helpfully to one who uses my terminology) termed the view that confirmation is probabilistic. There are however in vogue many objections to this view. Some of these we have already considered because they were aimed directly against individual axioms of the calculus. Others which deny resulting theorems (and thereby deny the axiom set of which the theorems are deductive consequences) will be considered in this chapter.

One of these objections arises from the paradox of ideal evidence. This is cited by Popper ([1] pp. 407f.) as an argument against what Popper calls ([1] p. 408) 'the subjective theory' of probability, but which he characterizes in such a way that the logical theory would be a 'subjective theory', and by Hooker and Stove [4] as an argument against the main part of Mackie's relevance criterion, that is my Principle C (see my p. 48). But, as far as I can see, it only tells against Principle C by telling against the definition of confirmation used in it, that e confirms h for background evidence k if and only if $P(h/e.k) > P(h/k)$. It can of course only 'tell against' a definition in the sense of indicating that the definition in question is not very useful and ought to be replaced by a better one. The paradox is as follows: Let p be 'the next throw of this die will be a six', q be 'the last 600,000 throws have resulted in 100,000 ± 40 throws of six', k be 'the die has six faces labelled respectively 1, 2, 3, 4, 5 and 6'. In that case on our intuitive understanding of probability, we would judge that $P(p/k) = \frac{1}{6}$ and $P(p/k.q) = \frac{1}{6}$. So on our definition q has neither confirmed and disconfirmed p, but has proved 'irrelevant' to it. But surely that is not so, the objection runs. Intuitively q is vitally relevant to the probability of p. The result of many tosses is vital evidence in assessing the probability of p.

But all that this argument shows is that 'irrelevant' is being used in a technical sense and this must be admitted. It was introduced by me, following other writers, as a technical term – see p. 4. Nevertheless q has not made it any more or less probable than it was before that the next throw will be a six (p), and so on our definition has proved irrelevant to p. When q is added to our evidence, there is still no reason for supposing any one face more likely to come up than any other. Why we feel that q is relevant to p is because q has raised enormously the probability of a different proposition from p, the proposition s 'the complex statistical probability of a throw of a six with the die is very close to $1/6$'. $P(s/k \cdot q) \gg P(s/k)$. Given k alone the die might well have been biased for or against six, and so the long run frequency of sixes could easily have been $\frac{1}{2}$ or 0. q virtually rules that out, and so enormously increases the probability of s being true.

The second objection which I will consider to the claim that propositions of epistemic probability conform to the axioms of the calculus is an interesting argument by Schlesinger [5] against Mackie's Principle 'C_1', which, with very slight differences which we can afford to ignore here, is the same as our Principle D. This, with qualifications – see p. 49 – which we can here ignore, states that e_1 confirms h better than e_2 does

if and only if $\dfrac{P(e_1/k \cdot h)}{P(e_1/k)} > \dfrac{P(e_2/k \cdot h)}{P(e_2/k)}$.

Schlesinger considers the case of a fairly simple hypothesis h_s which (with background evidence k) entails observation-reports e_2 and $\sim e_1$, and a complex hypothesis h_c which (with background evidence k) entails both e_2 and e_1 (while no hypothesis more simple than h_c entails both e_2 and e_1). k does not contain any evidence which affects the probability of these hypotheses, and so may be regarded as mere analytic evidence. In that case e_1 confirms h_c more than e_2 does. By Principle D this will be so if and only

if $\dfrac{P(e_1/k \cdot h_c)}{P(e_1/k)} > \dfrac{P(e_2/k \cdot h_c)}{P(e_2/k)}$. Now $P(e_1/k \cdot h_c) = P(e_2/k \cdot h_c) = 1$. So by

Principle D e_1 confirms h_c more than e_2 does if and only if $P(e_1/k) < P(e_2/k)$. Yet Schlesinger suggests that in this situation there is no reason to suppose that $P(e_1/k) < P(e_2/k)$ and that $P(e_1/k)$ may well equal $P(e_2/k)$, in which case the principle would not hold. Schlesinger considers a plausible argument against this suggestion: since h_s is simpler than h_c and, as I have claimed earlier and will argue more fully in Chapter VIII, greater simplicity normally means greater intrinsic probability, then necessarily $P(h_s/k) > P(h_c/k)$; yet both h_s and h_c (with k) entail e_2 while h_s (with k) entails $\sim e_1$ and h_c (with k) entails e_1; since both hypotheses entail e_2 while only one, and that in advance of observation the less probable one, entails e_1, e_1 is less to be expected than e_2 and so necessarily $P(e_1/k) < P(e_2/k)$. This argument would save the principle. Schlesinger however has a counter-argument. He claims that although h_s is simpler than h_c, $P(h_s/k)$ and $P(h_c/k)$ may both 'have no finite value'. This would be so, he argues, if there were an infinite number of theories equally simple with h_s, and so the probability that one of these theories was true was 'divided equally among' all the theories. For in that case, he claims, each of the theories including h_s would have 'no finite' probability. Clearly $P(h_c/k)$ is not greater than $P(h_s/k)$ and so it too would have 'no finite' probability. This point must be admitted for the reasons which I gave in Chapter V, if we suppose that the probabilities in question conform to the axioms of the calculus. Schlesinger goes on from there to claim that in consequence the fact that h_s entails e_2 and $\sim e_1$ while h_c entails both e_2 and e_1 (all given k) is no reason for supposing that $P(e_1/k) < P(e_2/k)$. This must be granted, and Schlesinger's argument is valuable in pointing out that in a case where an infinite number of theories are equally simple with each other, it cannot be the case that any of them has greater intrinsic probability than a less simple theory.

However, in his use of this point to provide a counter-example to Principle D Schlesinger has forgotten one of the conditions in his own example – that all theories simpler than h_c are incompatible with e_1, but not with e_2. This condition is important for Schlesinger, for if all of the infinite number of other theories equally simple with h_s predicted e_1, it seems highly doubtful to say the least, whether e_1 confirms h_c more than e_2 does. But now, given that all theories simpler than h_c are incompatible with e_1 but not with e_2, then even if the intrinsic probabilities of h_c and of each of the theories simpler than h_c have 'no finite value' there is nothing incompatible with the axioms of the calculus (unless we misguidingly add to them the Axiom III* set out on p. 72) in supposing that the intrinsic probability that one of the theories simpler than h_c is true (that is, that the disjunction of those theories is true) has a finite value greater than 0. Let us call the proposition that a theory simpler than h_c is true 't'. Then intuitively $P(t/k) > 0$, and this supposition is not incompatible with the axioms of the calculus. Yet, Schlesinger assumed, t entails $\sim e_1$. On the information given e_2 is not incompatible with any known theory. Hence surely $P(e_2/k) > P(e_1/k)$ whence it follows that Principle D gives the correct result for the cited example.[1]

The third objection which I shall consider to the applicability of the probability calculus concerns the claim of Principle E (p. 51) that $P(h_2/e)$ $\geqslant P(h_1/e)$ if h_1 entails h_2. One writer who has recently produced an argument apparently directed against this principle is R. Harré ([6] p. 170). Harré produces a number of apparent counter-examples to the principle. To take the simplest of these (and ignoring as not important for this issue, a slight qualification which he adds to the example), he claims that 'if h is Galilean mechanics, and j is Kepler's planetary law, then Newton's cosmology, which might be said to be the conjunction of h and j, is much more highly confirmed by the evidence e, the facts about motion of bodies known to Kepler and known to Galileo, than is either taken separately'.

It is not quite clear to me from his earlier discussions whether Harré understands 'is more highly confirmed by' in the sense of 'is more probable given', but I think that he does, and if this is so, this passage constitutes an argument against my Principle E.

However, if it is taken in this sense, Harré's claim is false. The true

(1) Schlesinger also produced a superficially plausible objection to Mackie's Principle 'C_2', which – see p. 50 n. 1 – is also a consequence of the axioms of the calculus. This is that, given k, e confirms h_1 better than it does h_2 if and only if $\dfrac{P(e/k \cdot h_1)}{P(e/k)} > \dfrac{P(e/k \cdot h_2)}{P(e/k)}$ which, given $P(e/k) \neq 0$, is to say, if and only if $P(e/k \cdot h_1) > P(e/k \cdot h_2)$. Against this Schlesinger argues as follows. Consider two hypotheses h_1 and h_2, both of which, together with k, entail the observation e and so $P(e/k \cdot h_1) = P(e/k \cdot h_2)$, yet h_1 is simple and h_2 complex. (They could be two of the curves marked on Figure 1 on p. 45). Yet surely in such a case Schlesinger urges, e confirms h_1 better than it confirms h_2.

That all depends on what we mean by 'confirms'. Certainly e makes h_1 more probable than h_2, as stated by my Principle A. In his discussion of Mackie's 'C_2' Schlesinger must be taking 'confirms' in this way. But if 'confirms' means what I have understood by it in this book – see p. 3 – Schlesinger's example does not tell against 'C_2'. It is only with this understanding of 'confirms' that Mackie's 'C_2' follows from the axioms of the calculus. Under this interpretation what Mackie's 'C_2' says is that e raises the probability of h_1 by a greater amount than it raises the probability of h_2 $\left(\dfrac{P(h_1/e \cdot k)}{P(h_1/k)} > \dfrac{P(h_2/e \cdot k)}{P(h_2/k)} \right)$ if and only if (given $P(e/k) \neq 0$) $P(e/k \cdot h_1) > P(e/k \cdot h_2)$. So if $P(e/k \cdot h_1) = P(e/k \cdot h_2)$, as in Schlesinger's example when e is a deductive consequence both of $(k \cdot h_1)$ and of $(k \cdot h_2)$, then the principle says that e raises the probabilities of both h_1 and h_2 above their previous values by the same proportion. This is quite compatible with, indeed entails, the hypothesis which was previously more probable remaining more probable.

situation is as follows. There are particular propositions of two kinds which were evidence for Newton's Theory – observations on the mechanical behaviour of bodies near the surface of the Earth, which I will term e_1, and observations of planetary positions which I will term e_2. For known initial conditions both e_1 and e_2 are predicted by Newton's Theory, which I will call T, and which, following Harré's simplification, we can treat as $(h.j)$. e_1 confirms T (given k), because it confirms h, which in turn confirms T. e_2 confirms T_1 (given k) because it confirms j, which in turn confirms T. (Given k) h is confirmed by e_1 and also less directly by e_2, because e_2 confirms j which confirms T, part of which is h. Similarly (given k) j is confirmed by e_2, and also less directly by e_1, because e_1 confirms h which confirms T, part of which is j. Now it may well be that $P(T/e_1.e_2.k) > P(h/e_1.k)$ and also $P(T/e_1.e_2.k) > P(j/e_2.k)$, and it seems to be this that Harré has in mind. But what is at stake is whether $P(T/e_1.e_2.k)$ is greater than $P(h/e_1.e_2.k)$ or than $P(j/e_1.e_2.k)$. And whatever the evidence surely T cannot be more likely to be true than h on the same evidence. For h is part of T. If T is true, h must be true. Hence in so far as $e_1.e_2.k$ confer a certain amount of probability on T, they thereby confer that amount of probability on whatever must be true if T is true, for example h. Yet h may derive more probability from $(e_1.e_2.k)$ than they confer on it via T (since $e_1.e_2.k$ may provide some reason for supposing h true which is not reason for supposing T true). So T, which is $(h.j)$ cannot be less probable than h on the same evidence.

My Principle E must not be confused with the consequence condition originally suggested by Hempel ([7], p. 31) under the name 'special consequence condition'. This says that if e confirms h_1, and h_1 entails h_2, then e confirms h_2. By 'confirms', it will be recalled, I understand 'renders more likely to be true', and Hempel had the same understanding as mine. Principle E says that if '$h_1 \supset h_2$' is analytic, then for all e and k, $P(h_2/e.k) \geqslant P(h_1/e.k)$. The consequence condition says that if '$h_1 \supset h_2$' is analytic, then for all e and k, if $P(h_1/e.k) > P(h_1/k)$, then $P(h_2/e.k) > P(h_2/k)$ – which is not the same.

The consequence condition seems plausible, yet the supposition that it holds for four propositions $P(h_1/e.k)$, $P(h_1/k)$, $P(h_2/e.k)$, $P(h_2/k)$ is, we will see, for some ascriptions of probability to them, inconsistent with the supposition that they obey the axioms of the probability calculus. Hence if the consequence condition holds generally that constitutes a further objection to the applicability of the axioms of the calculus.

We will take as an example to bring out the incompatibility of the consequence condition with the axioms of the probability calculus, a case when h_1 and h_2 are conjunctions of particular propositions. Let p and $\sim p$, q and $\sim q$ be reports of results of some experiment, each member of each pair being equally probable, both on background evidence k and on evidence of the result of the other experiment (e.g. k could be the information that a coin had been tossed twice, p could be the report that the coin landed heads on the first occasion, and q the report that the coin landed heads on the second occasion). Then

$$P(p/k) = P(q/k) = P(p/k . q) = P(q/k . p) = \tfrac{1}{2}.$$

Let h_1 be '$p . q$', h_2 be '$p \vee q$'.

Then, since by Axiom IV of the calculus

$$P(p . q/k) = P(q/k) \times P(p/k . q),$$
$$P(h_1/k) \quad = \tfrac{1}{4}.$$

Since by Theorem 4 (p. 41) of the calculus $P(p \vee q/k) = P(p/k) + P(q/k) - P(p . q/k)$.

$$P(h_2/k) = \tfrac{3}{4}.$$

Now let us be given new evidence e that '$(p . q) \vee (\sim p . \sim q)$' (e.g. that the coin landed the same way up on both occasions), then, by Theorem 4, $P(p/k . q . e) = 1$.

By Axiom III $P(e/k) = P(p . q/k) + P(\sim p . \sim q/k)$. We know that $P(p . q/k) = \tfrac{1}{4}$. We can work out, using various axioms and theorems, that $P(\sim p . \sim q/k) = \tfrac{1}{4}$. Hence $P(e/k) = \tfrac{1}{2}$.

By Axiom IV

$$P(h_1/e . k) = \frac{P(h_1 . e/k)}{P(e/k)}$$

and

$$P(h_2/e . k) = \frac{P(h_2 . e/k)}{P(e/k)}$$

$$h_1 \equiv (h_1 . e) \equiv (h_2 . e)$$

Hence by Axiom V

$$P(h_1/e . k) = P(h_2/e . k) = 2P(h_1/k) = \tfrac{1}{2}$$

So $P(h_1/e . k) > P(h_1/k)$, and e confirms h_1

$P(h_2/e . k) < P(h_2/k)$, and e disconfirms h_2.

Yet h_1 entails h_2. So evidence e confirms h_1 while it disconfirms h_2, although h_1 entails h_2 – which violates the consequence condition. This shows that *either* we cannot ascribe probabilities such that

$$P(p/k) = P(q/k) = P(p/k . q) = P(q/k . p) = \tfrac{1}{2}$$

or the axioms of the calculus do not apply in such a case *or* the consequence condition is false. Yet surely we can ascribe probabilities as above.
It is highly plausible that these are the probabilities in the cited coin-
tossing experiment.

So perhaps the axioms are at fault. Yet in the example cited the
results yielded by the axioms are highly plausible. With the cited interpretation for the coin-tossing experiment, undoubtedly, given k, h_2 is
more probable than h_1. Undoubtedly too e makes h_1 more probable
than it was before. There are four possible results of the experiment,
surely each given k equally probable $-(p . q)$, $(p . \sim q)$, $(\sim p . q)$, and
$(\sim p . \sim q)$. All that e does is to rule out two possible results, and leaves
h_1, that is $(p . q)$, as one of two equally probable results instead of one
of four equally probable results. And if we reflect we can see also that
e makes h_2 less probable than before. Without e there were four possible
results of the experiment, each equally probable, of which three constituted the fulfilment of h_2. Yet given e there are only two possible results,
both equally probable, of which only one constitutes fulfilment of h_2.
I conclude that the consequence condition is at fault.

Certainly the consequence condition *seems* plausible. As Mackie writes:
'In general the point of confirming hypotheses is to be able to draw predictions and other inferences from them, and we want some reliance on
these inferences to be justified by whatever evidence has confirmed the
hypothesis' ([9] pp. 35f). True, the criterion seems plausible, but surely
the example brings out that it is not a necessary truth. In that example
$(e . k)$ does not render h_2 more likely to be true than does k alone, and
so e does not confirm h_2, given k. No doubt usually if h_1 entails h_2, and
so for some proposition r, $(h_1 \equiv (h_2 . r))$, one expects that if $P(h_2 . r/e . k)$
$> P(h_2 . r/k)$, then $P(h_2/e . k) > P(h_2/k)$. But we have shown that this is
not a necessary truth.

Hence this fourth objection to the axioms of the calculus does not
work.

In the important Section 8 of his paper [7] Hempel referred to other
'conditions of adequacy' which, he claimed, must be satisfied by any
true account of confirmation. Since we have just been discussing one of
Hempel's conditions, it will be useful to comment here on the others.
(8.1) is the entailment condition – 'any sentence which is entailed by an
observation report is confirmed by it'. (I shall ignore difficulties raised –
see my p. 2 – by Hempel using 'sentence' instead of 'proposition'.) Now
it follows from Theorem 1 (p. 40) that if r entails q, then for all k,
$P(q/r . k) = 1$. But r confirms q if and only if it raises the probability of

q, that is if and only if $P(q/r . k) > P(q/k)$. So if $P(q/r . k) = 1$, r will confirm q if and only if $P(q/k) < 1$. By Theorem 1 again, $P(q/k) < 1$ only if k does not entail q. So the entailment condition does not hold if the background evidence already entails the hypothesis or if for any other reason the probability of the hypothesis on the background evidence k is 1. (For example the hypothesis q could be the negation of a hypothesis h which had a probability of 0 on k, even though k did not entail $\sim h$. Possible such cases were considered in the last chapter.)

(8.2) is the consequence condition, a more general form of (8.21), the 'special consequence condition' discussed above. Arguments against the latter tell also against the former. (8.22) is Hempel's 'equivalence condition', which I called in Chapter IV the c-condition. Hempel argues for it here that it is a deductive consequence of the special consequence condition. This argument is no use since we have shown the special consequence condition to be false. However I showed in Chapter IV that the c-condition was entailed by the hypothesis equivalence condition. I argued that analysis of the ordinary use of probability concepts does not provide a clear answer about whether the two equivalence conditions (our Axioms V and VI) are involved in our ordinary understanding of probability. If we wish to have a precise concept of probability, we need to make a decision whether or not to adopt these conditions, and I gave in Chapter IV some reason for doing so.

Hempel's next condition is his (8.3), consistency condition – 'every logically consistent observation report is logically compatible with the class of all the hypotheses which it confirms'. From this follows (8.31), 'unless an observation report is self-contradictory, it does not confirm any hypothesis with which it is not logically compatible', and (8.32), 'unless an observation report is self-contradictory it does not confirm any hypotheses which contradict each other'. (8.31) follows from our Theorem 1 (which depends on the evidence equivalence condition for its proof). By Theorem 1 (p. 40) if '$p \supset q$' is analytically true, $P(q/p) = 1$. Hence, by Theorem 2, $P(\sim q/p) = 0$. So if p and $\sim q$ are not logically compatible, i.e. are inconsistent (which will be the case if '$p \supset q$' is analytic), $P(\sim q/p) = 0$, and so for any k $P(\sim q/p . k) = 0$. Hence, since by Axiom I 0 is the minimum value of a probability, $P(\sim q/p . k)$ cannot be greater than $P(\sim q/k)$. Hence p cannot confirm $\sim q$ if p is not logically compatible with $\sim q$. (The possibility of 'p' being self-contradictory does not arise on the assumption which I make throughout that propositions describe logically possible states of affairs.)

(8.32), 'unless an observation report is self-contradictory, it does not confirm any hypotheses which contradict each other', however seems obviously false, quite independently of whatever view is taken about the equivalence conditions. Many examples discussed in earlier chapters provide counter-instances to it. For example, the hypothesis h_1 that variables x and y are related by the equation $x = (1 \pm r) y$, and h_2 that variables x and y are related by the equation $x = (1 \pm r)\{y + (y - 1)(y - 2)\}$ where r is a small constant, are both confirmed by the observation report that when y had to a high degree of approximation the value 1, x also had to a high degree of approximation the value 1. Clearly, in each case finding that the equation was satisfied in some instance made the hypothesis more probable. Yet the hypotheses are inconsistent; they contradict each other, since they make inconsistent predictions – h_1 predicts that when $y \simeq 3$, $x \simeq 3$; h_2 predicts that when $y \simeq 3$, $x \simeq 5$. Oddly, Hempel is aware of this objection, but fails to find it a conclusive one. From the falsity of (8.31) there follows the falsity of (8.3).

Also in this section (8) Hempel discusses but rejects a condition which is also considered by others, the converse consequence condition. This states that if e confirms h_1, it also confirms any hypothesis h_2 of which h_1 is a deductive consequence. Hempel rejects the converse consequence condition on the grounds that the consequence condition is true, and that if the converse consequence condition were also true, any evidence would confirm any proposition whatsoever – which is absurd. For, Hempel claims that e always confirms $e^{(1)}$ and so by the converse consequence condition for any h, e would confirm any proposition $(e \cdot h)$; and by the consequence condition whatever confirms $(e \cdot h)$ confirms h; hence any evidence e would confirm any hypothesis h. We have rejected the consequence condition, and so we cannot use Hempel's argument against the converse consequence condition. However, other arguments can be used to show the converse consequence condition to be false. Suppose our evidence e to be that very many ravens have been observed and found to be black and very many swans observed and found to be white. (For background evidence of most kinds) e confirms h_1 'all ravens are black'. h_1 is a deductive consequence of h_2 'all birds are black'. But e certainly does not confirm h_2; indeed it is incompatible with it. I conclude that

(1) Hempel argues from the entailment condition that e always confirms e. But as we saw a few pages back in discussing this condition, it holds subject to the proviso that the probability of e on background evidence alone is not already 1. We can add this assumption to Hempel's argument which, given the consequence condition, then shows that any evidence would confirm any hypothesis which did not already have a probability of 1 on background evidence alone, if the converse consequence condition were true.

G

the converse consequence condition is not a necessary truth. True there are many cases where it holds. Observations *e* of the passage of light near the sun confirmed the proposition h_1 that light passing close to a massive body is bent towards it in conformity with a certain equation, and so confirmed h_2, Einstein's General Theory of Relativity, of which h_1 is a deductive consequence. But the principle does not always hold.

Discussion of Hempel's Section 8 was prompted by consideration of his consequence condition, which has been used as an argument against the applicability of the axioms of the probability calculus. I pass to consider a fifth argument against this. It has been argued by L. Jonathan Cohen that not merely is the 'support' given by evidence *e* to hypothesis *h*, which he symbolizes by $s(h/e)$, not equal to $P(h/e)$, but that it is no mathematical function of $P(h/e)$, $P(e/h)$, $P(e)$, $P(h)$, when all these are probabilities which obey the mathematical axioms of probability. Cohen's argument is stated in his book *The Implications of Induction* [10] and is based on principles of evidential support first stated in two earlier articles.

Cohen's conclusion is derived from two principles of evidential 'support' which are alleged to be intuitively obvious. The first is a standard equivalence principle which puts as a principle for 'support' what is claimed for probability by my Axioms V and VI:

(1) For any propositions *e, f, h,* and *i*, if *e* is equivalent to *f*, and *h* to *i*, according to some non-contingent assumptions, such as the laws of logic or mathematics, then $s(h,e) = s(i,f)$.

The other principle which Cohen calls the instantial comparability principle, is novel. It runs as follows:

(3) For any e, e^1, p, p^1, u and u^1, if u and u^1 are generalizations, p and p^1 are just substitution instances of u and u^1 respectively, and e and e^1 are conjunctions of existentially quantified propositions, each of which predicates something of some unspecified individual element in the domain of discourse, then $s(u,e) > s(u^1,e^1)$ if and only if $s(p,e) > s(p^1,e^1)$.

Cohen adds ([10] p. 20) that 'by "just a substitution instance" is meant that all the relevant information given by p and p^1 is that the elements to which they refer satisfy u and u^1 respectively. Their modes of referring to these elements add no further relevant information.' The instantial comparability principle thus states that one generalization, 'all *A*'s are *B*', has a higher degree of support than another, 'all *C*'s are *D*', if and only if each prediction of the former that a particular *A* will be *B* has a higher degree of support than each prediction of the latter that a particular *C* will be *D* (given that nothing else is known about the *A* except that it is an A and about the C except that it is a C). Cohen assumes in his principles that evidence statements report, in his terms, 'the results of tests against relevant variables', by which he means that they report the results of attempts to test whether the hypothesis holds under circumstances which we have reason to believe might make a difference to whether or not it held.

I shall not sketch in detail all the stages of Cohen's argument to show the non-probabilistic character of the support function, but its tendency can easily be shown by an example illustrating its first stage. Let u^1 be 'all ravens are black'; u^n be 'all n-tuples of ravens are n-tuples of black birds', and p^i be 'the i-th raven to be observed will be black'. Now let us assume that there are at least n ravens. Then u^1 will be equivalent to u^n. Hence by the equivalence principle the support given by some evidence e to u^1 will be the same as the support given by the same evidence to u^n. But it follows from the instantial comparability principle that if the support given by certain evidence to two generalizations is the same, then the support given by that evidence to instances of those generalizations is the same. Hence the evidence e which supports an instance of u^1, e.g. p^1, 'the first raven to be observed will be black' gives exactly as much support to an instance of u^n, i.e. the conjunction $p^1 . p^2 \ldots p^n$, 'the next n ravens observed will be black'. Put symbolically, $s(p^1,e) = s(p^1 . p^2 \ldots p^n, e)$. Yet by Principle E (p. 51) derivable from the probability calculus $P(p^1/e) > P(p^1 . p^2 \ldots p^n/e)$ unless $P(\sim p^1 . p^2 \ldots p^n/e) = 0$, which latter will not normally be the case. Hence $s(h,e)$ does not equal $P(h/e)$ if $P(h/e)$ conforms to the axioms of the probability calculus. More elaborate arguments are used to show that $s(h,e)$ is no function of $P(h/e)$, $P(e/h)$, $P(e)$, and $P(h)$ if these conform to the axioms of the probability calculus. Cohen's arguments are rigorous and I have no wish to challenge them. What I do wish to do is to call into question his premises which lead to these results.

Clearly the correctness of the premises depends on what Cohen understands by 'support'. He is curiously brief about this. He writes ([10] p. 6) that 'instances of at least one type of inductive support are familiar now to everyone and frequently described, viz. when a hypothesis about the solution of some problem in natural science has been tested experimentally and the report of this evidence is said to support the hypothesis'. This remark, and another remark ([10] p. 7) when he seems to be analysing the same concept as Carnap is analysing under the name of 'confirmation', leads the reader to suspect that Cohen understands by the 'support' given to a hypothesis what Carnap (see my p. 4) understands by its quantitative 'confirmation', that is its epistemic probability in our sense. If this is what Cohen does mean by 'support', then, I shall show, his instantial comparability principle is false, and his anti-probabilistic conclusion unjustified. Yet only if Cohen's 'support' is taken as 'epistemi[c] probability' can his conclusion be regarded as showing that the axioms of the probability calculus are not applicable in some area, for otherwise it does not show that epistemic probability is not subject to the axioms of the probability calculus, but only that something else is not.[1] So henceforward I will understand his 'support' as 'epistemic probability' in my sense.

In that case Cohen's 'equivalence principle' follows from our two equivalence conditions, for which I produced some argument in Chapter IV. On the assumption that this principle is correct, let us investigate the novel principle – the instantial comparability principle – understanding Cohen's 'support' as 'epistemic probability'.

Suspicion is cast on this principle by the counter-intuitive character, under the stated interpretation, of some of the consequences which Cohen draws from the conjunction of it and the equivalence principle. One of these consequences is that stated above that the support given by any evidence (which reports the results of tests against relevant variables) to 'the first raven to be observed will be black' will be the same as the support given for any n ($n \leqslant$ the number of ravens in the universe), to 'the next n ravens to be observed will be black'. This being taken to imply that evidence which renders it likely to some extent that the next raven to be observed will be black renders it just as likely that the next trillion ravens will be black, is clearly absurd. If we have observed six ravens in relevantly varied conditions and five of them are black and one not black, then this evidence renders it quite likely that the next raven will be black, but highly unlikely that the next trillion ravens will be black. Equally counter-intuitive under the stated interpretation are other principles which Cohen derives less formally in his system, e.g. ([10] p. 63) where u_i denotes a generalization,

for any u_1, u_2 and e, if $s(u_1,e) \geqslant s(u_2,e)$, then $s(u_1 . u_2,e) = s(u_2,e)$,

and ([10] p. 63) when p_i is a substitution instance of u_i

(9) if $s(p_1,e) \geqslant s(p_2,e)$ then $s(p_1 . p_2,e) = s(p_2,e)$.

Thus if p_1 is 'the first swan to be observed will be white', p_2 is 'the first raven to be observed will be black', and both are fairly weakly supported

---

(1) Mr Cohen has made clear to me in correspondence that he does not understand by 'support' 'epistemic probability' in my sense. However, I include this passage in case any reader of his does understand his 'support' in this sense.

by the evidence, although p_1 is more strongly supported than is p_2, principle (9) tells us that $(p_1 . p_2)$ is as well supported as p_2. But if 'supported' means 'rendered likely to be true', this is absurd. It is more likely that the first raven will be black than that both the first raven will be black and the first swan white.

Once we have been led to regard the principle, under the interpretation being considered, with suspicion, it is easy to find counter-instances to it. Let e be evidence that millions of ravens have been examined under relevantly varied conditions and all found to be black apart from one found to be white. Let e^1 be evidence that one raven has been examined and found to be black. $u = u^1 =$ 'all ravens are black' then $s(u,e) = 0 <$ $s(u^1,e^1)$. But a particular prediction that the next raven will be black $(p = p^1)$ is surely better supported on evidence e than on evidence e^1. $s(p,e) > s(p^1,e^1)$. For another example, let $u =$ 'all swans are white', $u^1 =$ 'all ravens are black', e be evidence that all of many swans observed under relevantly varied conditions have been non-white, e^1 be evidence that half of the many ravens under relevantly varied conditions have been black and half non-black. Then $s(u,e) = s(u^1,e^1) = 0$. Yet $s(p,e) < s(p^1,e^1)$. Under the interpretation of 'support' as 'epistemic probability' the instantial comparability principle is clearly false. Hence Cohen's argument does not show that the axioms of the probability calculus do not govern epistemic probability. The true significance of Cohen's system will be brought out in Chapter XIII.

Bibliography

Popper's arguments against the logical theory of probability and against the view that confirmation is probabilistic, including the argument discussed, are to be found in:

[1] KARL R. POPPER *The Logic of Scientific Discovery*, London, 1959, Appendix *ix.

For counter-arguments to arguments of Popper not discussed in this chapter, see:

[2] L. JONATHAN COHEN 'What has Confirmation to do with Probabilities?' *Mind*, 1966, **75**, 463–81. See especially pp. 466–8.

[3] PATRICIA BAILLIE 'That Confirmation may yet be a Probability' *British Journal for the Philisophy of Science*, 1969, **20**, 41–51.

For the paradox of ideal evidence see also:

[4] C. A. HOOKER and D. STOVE 'Relevance and the Ravens' *ibid.*, 1968, **18**, 305-15. See pp. 308-11.

For Schlesinger's argument against Principle D see:

[5] G. SCHLESINGER 'On Irrelevant Criteria of Confirmation' *ibid.*, 1970, **21**, 282-7.

For Harré's argument against my Principle E and other arguments against the probabilistic character of confirmation, see:

[6] R. HARRÉ *The Principles of Scientific Thinking*, London, 1970, Chapter 6.

For the consequence condition and other confirmation conditions suggested by Hempel, see:

[7] C. G. HEMPEL 'Studies in the Logic of Confirmation' *Mind*, 1945, **54**, 1-26 and 97-121; reprinted with an additional postscript in *Aspects of Scientific Explanation*, London, 1965, pp. 3-51. My reference is to the latter version.

For critical assessment of Hempel's conditions along lines similar to mine see:

[8] RUDOLF CARNAP *The Logical Foundations of Probability*, 2nd edition, London, 1962, pp. 468-78.

For Mackie's treatment of the consequence condition and of Principle D see:

[9] J. L. MACKIE 'The Relevance Criterion of Confirmation' *British Journal for the Philosophy of Science*, 1969, **20**, 27-40.

For Cohen's theory of induction, see:

[10] L. JONATHAN COHEN *The Implications of Induction*, London, 1970.

CHAPTER VII

Simplicity

So far in the book, after an introductory discussion in Chapters I and II about the different kinds of probability, we have been concerned mainly with whether the axioms and resultant theorems of the probability calculus, taken as a calculus of epistemic probability, state true propositions. Our main concern, that is, has been with the logical relations between such expressions as $P(h/e)$, $P(h)$, $P(e/h)$, and $P(e)$. Those most important for our purposes I stated in five principles in Chapter III and subsequently defended against objections. It is now time to consider how quantitative or comparative values are given to such expressions. In this present chapter I consider how the intrinsic probability of nomological propositions is to be compared (given that for reasons given in Chapter III quantitative values cannot in general be given to such probabilities). In Chapter V I have defended the claim that certain nomological propositions have zero intrinsic probability, but that most universal nomological propositions have a probability greater than zero. We must now ask what features of two commensurable universal nomological propositions h_1 and h_2 make $P(h_1)$ greater than, equal to, or less than $P(h_2)$. The principles which I shall set forward in this chapter for determining that one hypothesis h_1 has greater intrinsic probability than another one h_2 are only meant to apply where the former one is not ruled out as non-projectible (and so $P(h_1) = 0$) by such considerations as those outlined at the end of Chapter V.

One principle which has immediate relevance to our problem is Principle E (see p. 51). Principle E states that if '$h_1 \supset h_2$' is analytically true, $P(h_2/e) \geqslant P(h_1/e)$; and that unless $P(\sim h_1 . h_2/e) = 0$, $P(h_2/e) > P(h_1/e)$. If we put e as a tautology it follows that if '$h_1 \supset h_2$' is analytically true, $P(h_2) \geqslant P(h_1)$ and unless $P(\sim h_1 . h_2) = 0$, $P(h_2) > P(h_1)$. Hence a wider nomological proposition has an intrinsic probability no greater than that of a narrower one. h_1, 'of physical necessity all metals dissolve in any acid', has an intrinsic probability equal to or less than h_2 'of physical–

necessity all iron dissolves in hydrochloric acid'. Now unless '$h_2 \supset h_1$' is also analytically true (and so h_1 and h_2 are logically equivalent) h_1 will be wider than h_2, and so it will entail nomological propositions other than h_2. Thus h_1 above also entails h_3, 'of physical necessity all silver dissolves in nitric acid'. Now in this case where h_1 is a universal nomological proposition wider than h_2, it seems highly implausible to suppose that $P(h_1/h_2) = 1$. Even if all iron dissolves in hydrochloric acid it is not certain that all metals dissolve in any acid. Hence, by Theorem 2 (p. 40), $P(\sim h_1/h_2) \neq 0$. By Axiom IV $P(\sim h_1 . h_2) = P(\sim h_1/h_2) \times P(h_2)$. Hence $P(\sim h_1 . h_2)$ will equal 0 if and only if $P(h_2) = 0$, that is if h_2 is not projectible. So, given that h_2 is projectible, when h_1 entails h_2 but h_2 does not entail h_1, $P(h_2) > P(h_1)$.

These considerations allow us to draw a wider conclusion, that even where h_1 does not entail h_2, but nevertheless has greater content than (is wider in scope than, says more about the world than) h_2, it is *as such* less likely to be true. If h_1 is 'all material bodies attract each other with forces proportional to mm'/r^2' (that is each body of mass m attracts each other body of mass m' at a distance from it of r with a force proportional to mm'/r^2) and h_2 is 'all iron dissolves in nitric acid', intuitively, h_1 says more about the world than h_2. Now if h_1 says more than h_2 does, a hypothesis h_3 can always be constructed by widening the claims of h_2, so that h_3 entails h_2 while being equally wide with h_1. By the argument of the previous paragraph (and subject to its qualification that the predicates involved in h_2 be projectible), $P(h_2) > P(h_3)$. Hence if h_1 is *either* equally probable with *or* less probable than h_3, h_1 will be less probable than h_2. h_1 will have to be more probable than h_3 even to equal h_2 in probability. *A priori*, it seems more likely that an hypothesis h_1 will be either equally probable with or less probable than an hypothesis h_3 than that it will be more probable than h_3. I conclude that in so far as an hypothesis is wider that is, as such, reason for supposing it less probable. I stress the words 'as such'. There may be reasons why a particular large-scale claim is more likely to be true than a certain small-scale claim. It may have greater simplicity in ways such as those to be described below.

The rules to be described henceforward are rules for assessing the comparative intrinsic probability of hypotheses of equal content or hypotheses whose relative content may be ignored. They describe factors which must be weighed against differences between hypotheses in respect of content in order to assess comparative intrinsic probability. In such a case where we ascribe greater intrinsic probability to one of two hypotheses of equal content, intrinsic probability seems to be a matter of simplicity, in that wide sense of the term in which it is used by philosophers of science, referred to earlier, which we shall illustrate by various examples. The rules which I shall give will be concerned solely with the relative simplicity of incompatible universal nomological propositions.

The simpler an hypothesis *is* in the ways to be described, given that it is projectible, the greater its intrinsic probability. In essence an hypothesis h_1 is simpler than another h_2 if it gives a more coherent picture of the world. But men's standards are very varied here; what seems to one man a coherent picture does not seem so to another; or even if all actual men agree that one hypothesis h_1 is simpler than h_2, it is often possible to conceive of men of a different culture who would make the opposite judgement, where we would not be prepared to call that judgement an irrational one, given their cultural environment. The purpose of this book is, I have stated, to draw out the criteria of epistemic probability used by men of our culture today. These are those criteria which are such that if men of our culture or any other culture use them, to that extent we judge them rational. Such criteria may be termed trans-cultural because we judge men of any culture rational to the extent to which they make judgements in accord with them. Hence, as we seek in this chapter to set out rules of comparative simplicity, we seek trans-cultural rules. However, as we shall see, it is not always possible to give such rules, because we may well judge men of another culture rational, even though they make different judgements of comparative simplicity from those which we would make. In such a case all we can do is to set out what I shall call this-cultural rules – that is, the rules which we of our culture follow in making judgements of the comparative simplicity of hypotheses.

So in this chapter I shall discuss a number of rules for judging comparative simplicity, rules which set out our ordinary standards in these matters. Some of these I shall judge trans-cultural and some merely this-cultural, but I feel very hesitant in apportioning the rules to the different categories. Others might have other views from mine about the judgements which we of our culture would make about the rationality of judgements of comparative simplicity implicit in the probability judgements of men of another culture. The only way finally to settle whether a rule was trans-cultural or merely this-cultural would be to take a survey of men's attitudes on this matter after the issues had been explained fully to them. The criteria which I shall outline will not allow us for all incompatible hypotheses of similar content h_1 and h_2 to conclude that h_1 is simpler than, as simple as, or less simple than h_2. But they will allow us to say this for some such hypotheses.

The simplest case of two incompatible universal nomological propositions arises where the propositions have the following form:

h_1:　all A's are B

h_2:　all A's are C

where B and C are interrelated by the following logical equivalences:

(x)　$\{Bx \equiv ((Cx \cdot Gx) \vee (Kx \cdot \sim Gx))\}$

(x)　$\{Cx \equiv ((Bx \cdot Gx) \vee (Jx \cdot \sim Gx))\}$

where 'C' and 'K' are contraries, that is of logical necessity cannot both apply to the same object at the same time, and so are 'B' and 'J'. Thus 'C' is defined such that everything which is G and B is C, and everything which is $\sim G$ and J is C. The two hypotheses make the same predictions about A's which are G but conflicting predictions about any A's which are $\sim G$ (the possible existence of which they do not rule out). h_1 says that they would be B; h_2 says that they would be C and so, by the equivalences, J, which is a contrary of B.

If we can construct a situation where the two hypotheses are equally accurate predictors of the evidence of observation e (and not incompatible with it), yet intuitively we can say that one hypothesis, say h_1, is rendered more probable by the evidence, then by Principle A (see p. 43 we can conclude that h_1 is intrinsically more probable than h_2. The two hypotheses will be equally accurate predictors, compatible with the evidence e, if we suppose that e is that a number of A's which are G have been observed and all of them found to be B and so C. So $P(e/h_1) = P(e/h_2)$. If in this situation we can say that $P(h_1/e) > P(h_2/e)$, we can conclude that $P(h_1) > P(h_2)$.

The most famous example of such a very simple conflict discussed by philosophers is the one put forward by Nelson Goodman in his 'new riddle of induction'. This was originally put forward in an article [1] in 1946 and more fully in Chapters 3 and 4 of his book *Fact, Fiction, and Forecast* [5] in 1953. Goodman supposes that we have observed a considerable number of emeralds and found them to be green (and he assume that there is no other relevant empirical evidence). On this basis we regard as moderately well confirmed by the evidence

h_1:　all emeralds are green.[1]

---

(1) Since the hypotheses up for consideration in this chapter are all nomological propositions, I shall not bother to include in the statement of each the expression 'of physical necessity'. This can be taken as read. In putting forward Goodman's hypotheses as examples of universal nomological propositions, I am assuming that he understood expressions of the form 'all A's are B' in this way.

h_1 will be in Goodman's terminology supported, non-violated and non-exhausted. An hypothesis of the form 'all A's are B' is supported, in Goodman's terminology, if an A has been observed to be B; it is violated if an A has been observed to be $\sim B$; it is exhausted if there are known to be no more A's. An A which is B I will term an instance of 'all A's are B'. Goodman termed an hypothesis which was 'adopted' when it was supported, non-violated, and non-exhausted, 'projected'. (If 'all A's are B' is projected, the predicate 'B' is said to be projected over the class of A's.) By 'adoption' Goodman means 'something like affirmation as sufficiently more credible than alternative hypotheses' ([5] p. 88). By 'credible' Goodman presumably means 'probable'. Then h_1 is projected if it is judged (much) more probable than other hypotheses about all A's which like the h_2 to be considered, make some predictions conflicting with those of h_1.[(2)] Goodman is seeking the rules governing 'valid projection' or 'projectibility', that is the rules which state when one such hypothesis is in fact (much) more probable than rivals.

Goodman now introduces 'grue' and 'bleen', which I will define as follows (I have altered Goodman's definitions slightly in a way adopted by others to make their point especially clear):

(x) $\{x$ is grue $\equiv ((x$ is green. x is at a time $<$ AD 2000) v $(x$ is blue. x is at a time $\geqslant$ AD 2000))$\}$

$(x)\{x$ is bleen $\equiv ((x$ is blue. x is at a time $<$ AD 2000) v $(x$ is green. x is at a time $\geqslant$ AD 2000))$\}$

Now since all emeralds so far observed which are green are also grue, the hypothesis
h_2: all emeralds are grue
will like h_1 be non-violated and non-exhausted, and be equally well supported by instances. Yet the two hypotheses 'all emeralds are green' and 'all emeralds are grue' while making the same predictions about the colours of emeralds before AD 2000 make different predictions about the colours of subsequent emeralds. One predicts that they will be green; the other predicts that they will be grue, and so blue. Which are we to adopt? Obviously h_1 'all emeralds are green'. But why? The task of confirmation theory is to bring out the principle which makes this the simpler

(2) Henceforward in order to simplify his problem I suppose Goodman to be concerned only with finding the more probable hypothesis, not the much more probable hypothesis. This simplification has no effect on any substantive issue.

and so more probable hypothesis (in Goodman's terminology, the more projectible hypothesis).

This conflict is an example of the pattern of conflict cited earlier, in which 'A' = 'emerald', 'B' = 'green', 'C' = 'grue', 'G' = 'is at a time <
AD 2000', 'J' = 'blue', 'K' = 'bleen'.[1] So $P(e/h_1 . k) = P(e/h_2 . k)$, when
k is a tautology and e reports that many emeralds have been observed and
found to be green. But $P(h_1/e . k) > P(h_2/e . k)$. So by Principle A,
$P(h_1/k) > P(h_2/k)$ (since for reasons given on p. 44) $P(e/k)$ and $P(e/h_1 . k)$
$\neq 0$).

The obvious rule to deal with Goodman's emerald example is that
provided at the end of Chapter V, that if 'B' is a qualitative predicate
and 'C' a positional predicate, we ought to project 'B' rather than 'C',
because nomological propositions containing positional predicates are
non-projectible absolutely, not merely less projectible than other

(1) In stating Goodman's example, I assumed, in order to give initially the
example of a very simple controversy, that the equivalences given above are
logical equivalences; that is that, as many writers on the subject have assumed,
it is part of the meaning of 'grue' that the equivalences hold. Goodman, for
reasons unconnected with this riddle, is very suspicious of alleged logical equivalences, and tends to treat the equivalences as synthetic, synthetic claims that
future objects will not be both grue and green. (For Goodman's understanding
of definition which shows this suspicion see [12]. An awkward consequence of
this understanding is noted later in this chapter – pp. 118f.) But in that case the
question arises of what evidence we could have for their truth. For some equivalences of the form of those given on p. 100 there could be background evidence
that they had been satisfied for objects other than A's which are $\sim G$. However
we cannot have any such evidence in this case. For no objects which are $\sim G$ can
yet have been observed, for '$\sim G$' means 'is at a time $\geqslant$ AD 2000'. If we suppose
that 'all emeralds are green' and 'all emeralds are grue' make conflicting predictions, this must be on logical grounds. (This claim is well made by Small in [10].)

This point is connected with a peculiarity of Goodman's example not noticed
by many participants in the controversy over Goodman's claims. The two hypotheses are cited and assumed by most writers to be hypotheses making predictions which could be observed to come off or not to come off. However the
Encyclopedia suggests that it is a defining characteristic of an emerald that it is
green. Hence it would seem that the hypothesis 'all emeralds are green' is analytically true and so 'all emeralds are grue' is analytically false. Hence the problem of which is more probable seems solved quite independently of considerations of the kind we are considering. Now we can easily construct a different
example of conflicting synthetic hypotheses, or we can continue to work with
Goodman's example on the assumption that it is not an analytic property but
only a contingent property of emeralds that they are green. In common with
many writers I shall do the latter. However it is a consequence of Goodman's
scepticism about logical truth that, for him, *all* 'definitions' are mere synthetic
statements, and so this feature of the definition of 'emerald' would not disturb
him.

hypotheses.[1] 'Grue' seems positional, and 'green' qualitative. Hence we ought to project 'all emeralds are green' and not 'all emeralds are grue'. In our terminology 'all emeralds are green' will be simpler and so, since its intrinsic probability is not zero, intrinsically more probable than 'all emeralds are grue'.

But Goodman claims we cannot show that 'green' is in itself qualitative and 'grue' in itself positional. He writes:

> True enough, if we start with 'blue' and 'green', then 'grue' and 'bleen' will be explained in terms of 'blue' and 'green' and a temporal term. But equally truly if we start with 'grue' and 'bleen', then 'blue' and 'green' will be explained in terms of 'grue' and 'bleen' and a temporal term. ([5] pp. 79f.)

Thus the following definitions of 'blue' and 'green' follow deductively from those given earlier of 'grue' and 'bleen':

(x) $\{x$ is green $\equiv ((x$ is grue. x is at a time $<$ AD 2000) v $(x$ is bleen. x is at a time $\geqslant$ AD 2000))$\}$

(x) $\{x$ is blue $\equiv ((x$ is bleen. x is at a time $<$ AD 2000) v $(x$ is grue. x is at a time $\geqslant$ AD 2000))$\}$

(The symmetry of interdefinability of 'grue' and 'green' is brought out by the definitions on p. 100.) So, according to Goodman, which of 'green' and 'grue' tells us about the relations of objects to a particular instant of time depends on which we define by which. He concludes:

> Thus qualitativeness is an entirely relative matter and does not by itself establish any dichotomy of predicates. ([5] p. 80)

Goodman's own solution to his riddle is that of two incompatible hypotheses, that hypothesis is projectible, the predicates occurring in which are better entrenched in the language. Very roughly, Goodman defines entrenchment so that a predicate 'P' is better entrenched in a language than a predicate 'Q' if it has occurred more often in hypotheses which have (rightly or wrongly) been projected previously. (A more precise account of entrenchment will be considered later in the chapter.)

> Plainly 'green' as a veteran of earlier and many more projections than 'grue' has the more impressive biography. The predicate 'green', we may say, is much better entrenched that the predicate 'grue'. ([5] p. 94)

(1) A *somewhat* similar suggestion was made by Carnap [2] in his reply to Goodman's original presentation of the issue [1]. Goodman replied to Carnap in [3], and Carnap made a rejoinder in [4]. I write *somewhat* similar because Carnap had a different understanding of 'projectibility' from ours (see [2] p. 146).

In the first edition of [5] Goodman stated three more detailed rules for determining projectibility; in the second edition he judged one of these superfluous and reduced the number to two; in a recent article [6] he judged yet another superfluous and put forward only one rule of projection. This is that (in my terminology) given two hypotheses h_1, 'all W's are B', and h_2, 'all Y's are C', which yield conflicting predictions and so are incompatible, if h_1 has a better entrenched antecedent predicate ('W') and a no less well entrenched consequent predicate ('B') than h_2, or has a better entrenched consequent predicate and a no less well entrenched antecedent predicate than h_2, then h_1 is to be projected in preference to h_2. (I shall henceforward consider only cases where the antecedent predicate is the same for the two conflicting hypotheses, but what I shall say can clearly be generalized to cases where the antecedent predicates differ.)

It is a consequence of Goodman's account that projectibility depends on degree of entrenchment, that 'all emeralds are grue' is merely *less* projectible than 'all emeralds are green', not absolutely non-projectible. Further, Goodman is committed to the view that if, as is possible, 'grue' were better entrenched than 'green' (because speakers of the language had projected 'grue' more often in the past) then 'all emeralds are grue' would be more probable intrinsically than 'all emeralds are green'; and so more probable on evidence predicted equally accurately by both hypotheses and compatible with them. This consequence seems mistaken – surely however misguided men had been in their past projections, observations of a number of green emeralds *in fact* makes it more likely that all emeralds are green than that they are all grue? The unsatisfactory consequence indicates that we should look for a different answer to the question what makes 'all emeralds are green' intrinsically more probable than 'all emeralds are grue'. Several writers have held that, despite the inter-definability of 'green' and 'grue' and members of similar pairs of predicates, it can be shown that predicates such as 'green' are essentially (i.e. quite apart from the history of the language in which they are used) qualitative, and predicates such as 'grue' essentially positional. Then the conflict can be resolved in the way suggested at the end of Chapter V, by saying that hypotheses containing positional predicates are not projectible at all (or by saying, at any rate, that they are less projectible than other hypotheses).

The best known of such attempts to solve the particular 'grue-green' conflict along these lines is that of Barker and Achinstein [7]. There was an answer opposing their solution by Goodman [8], and a defence of

Goodman by Ullian [9]. Believing the basic claim of Barker and Achinstein to be correct, I will give my own defence of it, which will lead to a rule of wider application.

There seems to be a crucial epistemological difference between 'grue' and 'green'. One can find out as certainly as can be found out at an instant whether or not an object is 'green' without being as a result in a position to state what is its relation to any other particular thing (whether instant of time, point of space, or physical object). To find out whether it is green, one has to find out whether it seems to most people in normal light to be the same colour as grass, trees, ivy, strawberry plant leaves, etc. The execution of such tests would put one in as certain a position as one can be at any instant to know whether an object is green. (This is not to deny that the object could be green even if the tests appear to have given negative results, and conversely; the observers might have lied; the light might not, despite appearances, have been normal etc. Nevertheless one can be nearer to certain knowledge by executing those tests than by executing any others – e.g. doing a chemical test on the object or asking observers who only saw it many days ago whether it then had the same colour as grass.) Yet having applied the tests which give one knowledge as near to certain as can be had at any instant whether the object is green, knowing the result of these tests would not put one in any position to know what was the time. To ascertain this one would have to perform further operations. I shall express this by saying that applying the primary tests for whether an object is green does not involve finding what is the time, i.e. the temporal relation of the object to a particular instant of time (e.g. the zero point of our calendar, the Birth of Christ). Further, it does not involve finding the relation of the object to any other particular thing, whether point of space, physical object or anything else.

However, it is different with 'grue'. There may be various ways of finding out whether an object is grue. Some observers from a tribe with special senses might be able to see straight off whether or not an object was grue. But the question is – do their observations give knowledge as certain as can be had? Surely not. The way to find out as certainly as can be found out whether an object is grue is to find out whether or not it is green and what the time is. The other method could (it is logically possible) have given results at variance with those of this method – in which case if its verdict was to be preferred, the logical equivalence stated on p. 101 would not have held. If you have found out whether or not an object is grue as certainly as you can, you will be in a position to state what is the time.

These epistemological differences seem to lie at the root of the distinction introduced at the end of Chapter V between positional and qualitative predicates. Henceforward we shall understand a predicate to be positional if in order to find out as certainly as can be found out whether or not it applies to an object we have to find out its relations to some other *particular* thing (e.g. a particular instant of time, point of space, or physical object). We shall understand by a predicate being qualitative that it is not positional in this sense. Once we understand 'positional' and 'qualitative' in this way, we can see that, despite the interdefinability of 'grue' and 'green', which is positional and which qualitative depends on the meaning of the terms. Hence we can defend against Goodman's objection our original claim put forward at the end of Chapter V that universal nomological propositions containing positional predicates are not projectible. This rule *seems* to me a transcultural rule.

The whole description of the conflict between h_1 'all emeralds are green' and h_2 'all emeralds are grue' might be objected to, on the grounds that we have a lot more evidence for determining which is more probable than mere observations of colours of emeralds, and in fact it is that further evidence which makes the former hypothesis more probable; and that the greater probability of h_1 on the evidence has nothing to do with either the positionality or the less well entrenched character of 'grue'. However, while it must be admitted that in fact we do have much more evidence, the effect of that evidence can only be assessed if we have some such criteria as those discussed for comparing hypotheses in respect of intrinsic probability. Thus we may suppose that we have a vast amount of background evidence about the behaviour of objects other than emeralds which makes probable H_1, 'objects of one kind do not all suddenly change their colour with the arrival of a new year', and so h_1, a deductive consequence of H_1. Yet whatever evidence we now have in 1972 it will be evidence about the behaviour of objects up to 1972. Hence H_1 is just as accurate a predictor as H_2, 'objects of one kind do not all suddenly change their colour with the arrival of a new year, except that emeralds change colour from green to blue at the arrival of AD 2000'. Observational evidence is equally well predicted by H_1 and by H_2. If H_1 is rendered more probable by that evidence it must be, in virtue of Principle A, because it is intrinsically more probable. We are still left with the problem of finding what makes an hypothesis like H_1 intrinsically more probable than an hypothesis like H_2, and such considerations as positionality or entrenchment are again relevant. Although we simplified the issue in supposing that our only evidence was of

emeralds observed to be green, the same kind of problem arises if we
suppose our evidence to be very much wider.

The 'grue-green' conflict described by Goodman is a special case of
the kind of conflict described on p. 100 in the respect that the differentiating factor 'G' here has reference to a particular individual, an individual instant of time (AD 2000). If we characterize the epistemological
difference between 'grue' and 'green' described above somewhat differently, we can generate a wider rule for dealing with some conflicts of the
type described on p. 100 where neither 'B' nor 'C' are positional predicates. It follows from the arguments above that in order to find out as
certainly as can be found out whether an object is grue, we have to find
out whether it is green; but that the converse does not hold. If finding
out whether or not 'C' applies to an object as certainly as can be found
out involves getting in a position to know whether or not 'B' applies to
it, but not conversely, then I shall term 'B' epistemologically superior
to 'C'. 'Green' is thus epistemologically superior to 'grue'. I suggest that
in all conflicts between incompatible hypotheses 'all A's are B' and 'all
A's are C', where 'B' is epistemologically superior to 'C', 'all A's are B'
is simpler and so intrinsically more probable than 'all A's are C' (and is
in consequence made more probable by evidence which both hypotheses
predict with equal accuracy). This rule is similar to one provided by
Blackburn in [11]. Like the earlier rule about hypotheses containing
positional predicates, it seems to me (just) a transcultural rule of comparative simplicity.

This rule, I suggest, operates whether or not the differentiating factor
G refers to a particular object. Only if it does will the epistemologically
inferior predicate be a positional one. An example of an epistemologically inferior predicate which is not positional is given by 'plan-black'
which is epistemologically inferior to 'black', when I define an object as
plan-black if and only if it is black and situated on a planet more than
80 million miles from its centre of revolution (e.g. on the Earth), *or* not-
black and not situated on such a planet.

To find out as certainly as we can whether an object is plan-black we
have to find out whether or not it is black (and is situated on a planet
more than 80 million miles from its centre of revolution) but to find out
for certain whether it is black we do not have to find out whether or not
it is plan-black. Intuitively 'all ravens are black' is more probable than
'all ravens are plan-black' on evidence that a number of ravens have been
observed on Earth and found to be black and so plan-black, and so by
Principle A is intrinsically more probable. This result is yielded by our

H

rule since the consequent predicate of the latter hypothesis is epistemologically inferior to that of the former. Consideration of similar examples
will, I suggest, make this rule even more plausible.

However if the argument at the end of Chapter V be accepted, there
appears to be between 'all emeralds are grue' and 'all ravens are plan-black'
the important difference that no evidence will confirm the former, while
evidence can confirm the latter. As far as the rules which I have so far put
forward go, 'all ravens are plan-black' is confirmed by finding of known
ravens that they are plan-black; but it is not rendered as probable as is
'all ravens are black' by finding of known ravens on Earth that they are
black and so plan-black. The plausibility of this result is brought out by
the fact that we can imagine circumstances in which 'all ravens are plan-
black' would be rendered substantially probable. This would be the case
if, for example, we had a vast amount of evidence from many planetary
systems showing that the colours of ravens, and perhaps of animals
generally, changed immediately they crossed an imaginary sphere with a
radius of 80 million miles centred on the centre of revolution. We should
certainly need a lot of evidence, for 'all ravens are plan-black' is an hypothesis of a radically different kind to that current in science. However
such evidence is conceivable. That supports the claim that 'all ravens are
plan-black' does not have zero intrinsic probability and so is confirmed
by finding of known ravens that they are plan-black.

If the arguments of Chapter V against universal nomological propositions containing positional predicates having zero intrinsic probability
be rejected, the greater intrinsic probability of 'all emeralds are green'
over 'all emeralds are grue' still follows from the wider rule of epistemological priority which I have put forward.

More substantial difficulties arise when the predicates 'B' and 'C' in
incompatible hypotheses 'all A's are B' and 'all A's are C' do not differ
in epistemological priority. There are two possible cases here. One possibility is that finding out (as certainly as can be found out) whether 'B'
applies to an object puts one in a position to know whether 'C' applies
and conversely. The other possibility is that finding out (as certainly as
can be found out) whether 'B' applies does not put one in a position to
know whether 'C' applies *and conversely.* In the former case I shall say
that 'B' and 'C' are epistemologically equivalent; in the latter case I shall
say that they are epistemologically independent.

A case of conflicting hypotheses whose consequent predicates are
epistemologically equivalent is provided by the curve-fitting problem.
Suppose that we are recording values of some variable Y which occur

for various values of another variable X, and that we have observed pairs of values x of X and y of Y (approximately) as follows: (1, 1), (2, 2), (3, 3), (4, 4) and (5, 5), where the second figure of each pair denotes the value of Y observed when the value of X was that recorded by the first figure of that pair. Then two equations, '$y = x$', and '$y = x + (x - 1)(x - 2)$-$(x - 3)(x - 4)(x - 5)$' predict the observations (with which they are compatible) with equal accuracy. To put the matter more formally, let h_1 and h_2 be as follows:

h_1: all values x of X are related to simultaneous values y of Y by the equation $y = x$;

h_2: all values x of X are related to simultaneous values y of Y by the equation $y = x + (x - 1)(x - 2)(x - 3)(x - 4)(x - 5)$.

Now these h_1 and h_2 are instances of hypotheses 'all A's are B' and 'all A's are C' related by the logical equivalences given on p. 100 for the following interpretations of the terms:

'A' = 'a value of X'
'B' = 'a value x of X related to the simultaneous value y of Y by the equation $y = x$'
'C' = 'a value x of X related to the simultaneous value y of Y by the equation $y = x + (x - 1)(x - 2)(x - 3)(x - 4)(x - 5)$'
'G' = '1, 2, 3, 4 or 5'
'K' = 'a value x of X related to the simultaneous value y of Y by the equation:

$$y = \frac{x(x - 1)(x - 2)(x - 3)(x - 4)(x - 5)}{(x - 1)(x - 2)(x - 3)(x - 4)(x - 5)}\text{'}^{(1)}$$

'J' = 'a value x of X related to the simultaneous value y of Y by the equation:

$$y = \frac{(y - x)\{x + (x - 1)(x - 2)(x - 3)(x - 4)(x - 5)\}}{(y - x)}\text{'}^{(2)}$$

To find out as certainly as can be found out whether or not 'B' applies involves us in finding out what is the value of Y for a simultaneous value of X, which puts us in a position to know whether or not 'C' applies, and conversely.

(1) Since division by 0 is not allowed in mathematics, this equation has no solution if $x = 1, 2, 3, 4$ or 5, and so if x is 1, 2, 3, 4 or 5 it cannot have the property K.
(2) Since division by 0 is not allowed in mathematics, this equation has no solution if $x = y$, and so if $x = y$, it cannot have the property J.

Attempts have been made to solve the curve-fitting problem in terms of some formula for measuring the relative mathematical complexity of equations used in rival nomological propositions. The proposition, the equations of which are mathematically less complex (or more simple), is then said to be simpler.

In *Scientific Inference* [15] and other works Harold Jeffreys proposed a solution to the curve-fitting problem as follows. (Outlines of a similar solution have been given by Kemeny [16].) He supposed that any equation of interest 'can be expressed as a differential equation of finite order and degree, in which the numerical coefficients are integers' ([15] p. 45). (A differential equation is one whose terms are either powers of the variables multiplied by a constant, e.g. $4x^3$, or derivatives of the variables with respect to each other multiplied by a constant, e.g. $2d^2y/dx^2$.) He argues that equations of other types in physics are hardly ever derived directly from observations and arise only at a theoretical level. Jeffreys's method for ranking equations of the former type has been stated somewhat differently in different publications, and interpretations of what Jeffreys really meant vary. But here is one reasonable interpretation by Mary Hesse of what Jeffreys 'essentially' proposed: 'The complexity of a law should be defined as the sum of the absolute values of the integers (degrees and derivative orders) contained in it, together with the number of its freely adjustable parameters. Thus $y^2 = ax^3$ would have complexity value 6; $d^2y/dx^2 + 2 \cdot 345\, y = 0$, complexity 4, and so on' [17].

Now whether or not equations other than differential equations have in the past been formulated on the basis of observations, clearly some transcendental equations (e.g. $y = e^x$) are simpler than some differential equations, and hence if theories were proposed with equations of the two types, there could be cases where the theory with the transcendental equation was intrinsically the more probable. Jeffreys does not provide rules for ranking such equations in respect of simplicity. Further, some counter-intuitive results emerge from his ranking. Thus '$y = x^{13}$' (complexity value 14) is on his ranking more complex than '$d^2y/dx^2 = ax + bx^2 + cx^3 + 3$' (complexity value 12).

However, it might be possible to amend Jeffreys's system or to produce a system of the same kind which was not open to these objections. The system would produce a ranking for equations of all kinds, including transcendental equations, and not yield counter-intuitive results. A perfect such system would not yield results more precise than those on which the vast majority of scientists would agree.

So we may be able to codify the current judgements of relative mathematical simplicity of equations which would be made by scientists of our

culture. However, there seems to be a case here where we would judge
men of other cultures with other scientific backgrounds as rational in
making different judgements of relative simplicity. For whether a scientific theory stating a mathematical correlation between various variables
is simpler than another one expressing a correlation between the same
variables would seem to depend on the notation in which it is expressed
and the way in which those who use the theory have come to understand
that notation.

Thus Jeffreys's system seems rational because it seems to embody at
any rate two plausible considerations. One is that a hypothesis h_1 is simpler than a hypothesis h_2 if h_2 contains all the terms which h_1 contains
and additional terms as well. Thus

let h_1 be '$x = y + 1$'

while h_2 is '$x = y + (y - 1)(y - 2)(y - 3) + 1$ $(= y^3 - 6y^2 + 16y - 5)$.

h_1 and h_2 will agree in predicting observations at $(2, 1)$, $(3, 2)$ and $(4, 3)$
yet disagree in other predictions. Yet intuitively we would say, and
Jeffreys's rules bring this out, that h_1 is simpler than and so has greater
intrinsic probability than h_2. However a system of notation can always
be chosen in which h_2 no longer differs from h_1 solely in containing more
terms; in fact a system can always be chosen in which h_1 differs from h_2
solely in having more terms. One just has to choose a system in which
what h_1 wants to say is awkward to express, while there is a peculiarly
simple way of expressing h_2.

Here is a system of notation in which the h_1 and h_2 of the last paragraph seem to have equal complexity. In it there occur only expressions
of two types: the sign of equality ('=') and expressions '$f(, , ,)$' in which
the four vacant spaces are occupied by positive real numbers (or zero) or
variable substitutes for them. We are allowed to add expressions of the
latter type, which is done by simply placing them next to each other,
but there is no equivalent in this notation to operations of subtraction,
multiplication or division. $f(n, x, a, b)$ is equivalent in our usual notation
to $ax^n - bx^{n-1}$ in the sense that, for any set of values of x, n, a, b, a user
of the novel notation judges it right to apply $f(n, x, a, b)$ in any circumstances where we judge it right to apply $ax^n - bx^{n-1}$ (e.g. he judges it right
to say that there were $f(1, 2, 1, 0)$ apples in a box where we judge it right
to say that there were $(1 \times 2^1) - (0 \times 2^0) = 2$ apples. However those who
use the notation do not think of it as having that meaning. Now in the
new notation h_1 will be represented by

$$f(1, x, 1, 0) = f(1, y, 1, 0) f(0, y, 1, 0)$$

while h_2 is represented by

$$f(1, x, 1, 0) = f(3, y, 1, 6) f(1, y, 16, 5)$$

The alleged superiority of h_1 by reason of it containing more terms has disappeared in this notation.

The other intuition behind Jeffreys's system is that some mathematical operations are essentially more complex than others, and that its use of a more complex operation ought to count against some formula when it is being given a complexity ranking. Thus, following in this tradition, one could claim that multiplication is an essentially more complex operation than addition, because to understand the former you have to understand the latter but not conversely. By a similar argument one could claim that raising to a power is essentially more complex than multiplication, logarithms more complex than powers etc. Now certainly to understand a power x^n as we understand it we have to understand it as representing x multiplied by itself n times. But there could be a community which had an expression x^n for which they could produce the value, which would be the same value as that which we ascribe to it for any given values of x and n, without understanding it in our way. For them raising x to the n-th power would seem a very natural operation on x in no need of explanation in terms of another operation. (For them it would be a mathematical *discovery* that x^n was equivalent to x multiplied by itself n times.) Such a community would not necessarily judge an equation with powers inferior in simplicity to one using only multiplication.

I conclude that our judgements about the relative mathematical simplicity of mathematical equations are dependent on the notation in which those equations are expressed and our familiarity with and understanding of that notation. A community using a different notation or having a different understanding of our notation would make different judgements of relative simplicity, and their judgements would seem to us in the light of their circumstances no less rational than ours. To some extent the difference of judgements about relative mathematical simplicity may be a matter of their having less knowledge of analytical truths than ourselves for example not knowing that x^n is x multiplied by itself n times. But even if they did have the same analytical knowledge as ourselves they could still make different judgements of simplicity – they might just regard x^n as a primitive expression, whatever analytical truths they knew about it. A system of rules for ranking formulae in respect of mathematical simplicity would thus seem to be merely a system of this-cultural rules.

In Chapter V I adduced an argument to show that all universal nomological propositions attributing to objects an exact numerical value of a measurement of a quantity which can vary continuously have zero intrinsic probability. If this argument be accepted, all the hypotheses of this character discussed in the last four pages must be read as containing an approximating device, before they can be compared in respect of intrinsic probability. Thus the equation in h_1 on p. 109 must be read as '$y = (1 \pm r)x$' and that in h_2 on p. 109 as '$y = (1 \pm r) \{x + (x - 1)(x - 2)\text{-} (x - 3)(x - 4)(x - 5)\}$', where r is some small constant. In general I suggest physicists are more interested in obtaining hypotheses of the right form rather than obtaining hypotheses with perfectly correct values of the constants; and so their interest lies in comparing hypotheses of the above type.

The considerations which I have adduced in the last few pages about the simplicity of an hypothesis depending on the familiarity of its terms and so being culturally relative, apply, I would suggest, generally to incompatible hypotheses 'all A's are B' and 'all A's are C' where 'B' and 'C' are epistemologically equivalent or epistemologically independent predicates. We can draw up lists of rules for determining which hypothesis would be judged, and rationally judged, simpler by men of our culture, yet we would regard men of other cultures as rational in making different judgements. The hypothesis containing that predicate which seems to provide a more natural description of objects is judged simpler. Yet which predicate seems to provide a more natural description of objects depends on the conceptual scheme with which we have grown up.

Let us give one or two more examples to make this claim plausible. A large number of ravens are observed, and found to be black, the same colour as present MA gowns of the University of Oxford, the oldest university of the English speaking world. Two hypotheses are proposed:

h_1: all ravens are black;

h_2: all ravens are of the same colour as Oxma gowns.

('Oxma gown' is defined as 'MA gown of the oldest university in the English speaking world'.) h_1 and h_2 are equally accurate in their predictions so far ($P(e/h_1 . k) = P(e/h_2 . k)$), but they make conflicting predictions about what will be the colour of ravens if the colour of Oxford MA gowns is changed. On the evidence h_1 is more probable than h_2 because h_1 is simpler than h_2. Why is h_1 simpler? Because 'B', 'black', is a more natural predicate, describes a property which in our conceptual scheme is a property more essential to ravens than 'C', 'has the same colour as Oxma gowns'. ('B' and 'C' are epistemologically independent; one can find out whether an object has the same colour as an Oxma gown by comparing them together without comparing them with standard objects to find out for certain what that colour is, and conversely.)[1] We cannot avoid the force of the argument in this and similar examples by claiming that in actual fact we have a lot of evidence (*e*) to show that 'black' is a more constant property of birds than 'of the same colour as Oxma gowns' (*H*), because, even if we do, as we saw on pp. 106f. it is only in virtue of the considerations which we have been discussing (viz. considerations of relative simplicity) that *e* gives any probability to *H*.

Here is another example. Chemists study the structure of water (identified by appearance, taste, and various physical tests such as its boiling point) and find by various experiments that all examined samples consist of molecules, each consisting of two atoms of hydrogen and one of oxygen, oxygen being the gas which the majority of animals of the environment take in from the atmosphere and which sustains life in them. Two hypotheses are proposed:

h_1: all molecules of water consist of two atoms of hydrogen and one atom of oxygen;

h_2: all molecules of water consist of two atoms of hydrogen and one atom of that gas from the atmosphere which the majority of animals of the environment take in from the atmosphere and which sustains life in them.

These two hypotheses agree in their predictions about the structure of molecules of water in our environment in which oxygen is that gas which the majority of animals take in and which sustains life in them ($P(e/h_1 . k) = P(e/h_2 . k)$). But for a planet on which the majority of animals are so made that they breathe in nitrogen which sustains life in them, the hypotheses disagree in their predictions. The '*B*' and '*C*' predicates are again epistemologically independent. We regard h_1 as simpler and so more probable on the observations made so far, but surely only because the concepts involved in it fit more naturally into our scientific conceptual system.

There does seem however at least one further rule which can be given for selecting from incompatible universal nomological propositions the one which is simpler and so intrinsically more probable, a rule which, like the rule of epistemological priority, is a transcultural rule, in the sense that we would judge rational a man of whatever culture in so far as he made judgements in accordance with it. This enables us to make certain judgements of comparative simplicity between hypotheses containing

(1) In this example, in terms of the equivalences of p. 100 '*B*' is 'black', '*C*' is 'has the same colour as Oxma gowns', '*G*' is 'belongs to a world where Oxma gowns are black', '*K*' is 'is black and does not have the same colour as Oxma gowns', '*J*' is 'is not black and has the same colour as Oxma gowns'.

predicates which are epistemologically equivalent or independent as well as between ones containing predicates which differ in epistemological priority.

We feel that 'grue', 'plan-black', etc. have an essentially *disjunctive* nature. Something is grue if and only if it is *either* (green and at a time < AD 2000) *or* (blue and at a time ≥ AD 2000). Green, it is true, can as we have seen be given a similar disjunctive definition. However we feel that 'grue' is *really* a disjunctive predicate while 'green' is not.

An objective characterization of disjunctiveness has been provided by Sanford [14]. Sanford begins by introducing the concept of a disjoint predicate. A predicate '*P*' is disjoint if and only if it can be 'partitioned' into two sub-predicates '*Q*' and '*R*' which have no borderline cases in common. (A borderline case of some predicate '*Q*' is an object which is just on the edge of being '*Q*'; which, if it is *Q* only just succeeds in being *Q*; if it is not *Q*, it is only just not *Q*. Thus a turquoise object is a border-line case of 'green'.) This means that:

   (i) Of logical necessity, every object which is *P* is either *Q* or *R*, but
       not both, i.e. $(x) (Px \equiv (Qx \lor Rx))$ and $(x) (Qx \supset {\sim}Rx)$;
  (ii) It is logically possible that there be cases and borderline cases of
       both '*Q*' and '*R*';
 (iii) Of logical necessity, any borderline case of '*Q*' is not a border-
       line case of '*R*'.

An example of a predicate which is disjoint, on Sanford's definition, is '*P*', 'is less than 5 feet tall or more than 6 feet tall'. '*P*' can be partitioned into '*Q*', 'is less than 5 feet tall' and '*R*', 'is more than 6 feet tall' so that these conditions hold:

   (i) Of logical necessity everything which is '*P*', 'less than 5 feet tall
       or more than 6 feet tall' is either '*Q*', 'less than 5 feet tall' or *R*,
       'more than 6 feet tall' and of logical necessity nothing is both *Q*
       and *R*;
  (ii) There can be objects which are *Q* and objects which are *R* and
       borderline cases of both, that is objects which are on the edge of
       being either, i.e. objects just about 5 feet tall or just about 6 feet
       tall;
 (iii) Any object just about 5 feet tall will not be an object just about
       6 feet tall.

Any predicate which is disjoint is, on Sanford's account, thereby an 'exclusively disjunctive predicate'. The spatial analogy of a disjoint predi-cate is a disjoint area such as Pakistan which 'consists of two provinces – West and East Pakistan – approximately 1,000 miles apart, separated by the Republic of India'. In contrast 'the area constituted by Colorado and Arizona is not disjoint, for the two states have one boundary point in common'. Such an area is said to be disconnected. By analogy Sanford defines a disconnected predicate 'P' as 'one which can be partitioned into two sub-predicates such that every borderline case of each sub-predicate is also a borderline case of the original predicate' ([14] pp. 164f.). On the definition cited earlier 'grue' will be a disconnected predicate. For it can be partitioned into 'Q', 'green, and at a time $<$ AD 2000' and 'R', 'blue and at a time $\geq$ AD 2000'. Every borderline case of 'Q' is a borderline case of 'P', and so is every borderline case of 'R'. Whereas green is, to all appearances, not a disconnected predicate. Although 'P', 'green', can be partitioned into 'Q', 'green at a time $\leq$ AD 2000' and 'R', 'green and at a time $\geq$ AD 2000', there will be borderline cases of 'Q', e.g. objects which are green around midnight on 31 December AD 1999 which are not borderline cases of 'P', 'green'. There is no doubt that they are 'P' green, but some doubt about whether it is AD 2000 or not, and so some doubt whether they are Q or R. Sanford considers ways in which it could be claimed that 'grue' was not disconnected, but we will not consider these, as we already have a rule from which it follows that 'grue' is not projectible. Both disjoint and disconnected predicates and also predicates of a kind which Sanford calls 'skew', are what Sanford calls exclusively disjunctive predicates.

Sanford defines a predicate 'P' as 'inclusively disjunctive' if it can be 'split' into sub-predicates 'Q' and 'R' which are jointly exhaustive of it yet not exclusive of each other (that is, anything which is P is either Q or R, and may be both), such that if anything is a borderline case of both 'Q', and 'R', and '$Q \cap R$' (an object is '$Q \cap R$' if it is both Q and R), it is necessarily the case that it is a borderline case of '$Q \cap \sim R$' and of '$R \cap \sim Q$'. Thus, 'P', 'red or hard' is inclusively disjunctive because it can be split into 'Q', 'red', and 'R', 'hard', such that anything which is P must be either Q or R and may be both; and any object which is a borderline case of 'red' and of 'hard' and of 'red and hard' is necessarily a borderline case of 'red and not hard' and of 'hard and not red'. We feel 'red or hard' to be an essentially disjunctive predicate. Sanford's rule brings this out.

He contrasts 'red or hard' with a predicate 'redange or yellange':

Let us understand 'redange' to apply to red things, orange things, and things which are borderline cases both of 'red' and of 'orange' but not of 'neither red nor orange'. And let us similarly understand 'yellange' to apply to yellow things, orange things, and things which are borderline cases both of 'yellow' and 'orange' but not of 'neither yellow nor

orange'. Now consider the predicate 'redange or yellange'. It applies
to colours in the spectrum between red and yellow inclusive. Although
it is broader than most of our colour terms, there is intuitively no
reason to call it a disjunctive predicate. Something which is a borderline case both of 'orange' and of 'brown' is a borderline case both of
'redange' and of 'yellange'. It is also a borderline case of the compound
predicate 'redange and yellange'. But it is not a borderline case of
either 'redange and not yellange' or 'yellange and not redange'.
([14] p. 167)

It will perhaps be useful to clarify this example by a map of the
colours. I mark by *a,* the object just referred to (Figure 2).

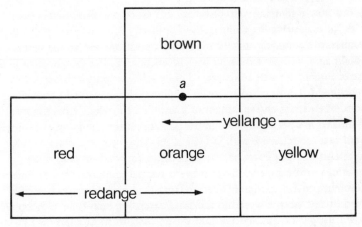

FIGURE 2

Exclusively disjunctive and inclusively disjunctive predicates are all,
on Sanford's account, disjunctive predicates. With this objective account
of disjunctiveness, we can now set forward the following plausible rule of
simplicity. h_1, 'all A's are B', is simpler than h_2, 'all A's are C', if 'B' is a
non-disjunctive predicate, while 'C' is a disjunctive predicate. This seems
a useful further rule of simplicity and, I suggest, a transcultural one.

However, although there are these transcultural rules for comparing
in respect of simplicity incompatible universal nomological propositions,
they do not seem able to deal with the vast majority of cases. They do not
for example deal with mathematical cases considered earlier. In such cases
one of the predicates does not seem to have some essential character such
as epistemological inferiority or disjunctiveness, which rules it inferior to

its rival. Its inferiority is, rather, a matter of its being clumsy or unnatural for a speaker of the language to use. My grounds for saying this is that it is possible for so many cases to imagine speakers of other languages preferring alternative hypotheses to those which we adopt; and our judging them, in the light of information about their history, rational to do so. In the vast majority of cases the greater simplicity of a theory consists in the greater familiarity of its concepts. In such cases we cannot really talk about one theory being simpler absolutely than another theory, but only simpler for this or that cultural group. It is hard to resist Goodman's general conclusion that we project and judge it rational to project the familiar – although we saw reason to reject it as an account of why we do not project 'all emeralds are grue'.

But how is familiarity to be measured? Goodman proposed certain rules for measuring the comparative familiarity of predicates and so the comparative simplicity of nomological propositions containing them. Goodman measures familiarity by 'entrenchment'. For Goodman, a term's entrenchment consists of its earned entrenchment plus its inherited entrenchment. Let us consider for the moment simply earned entrenchment. The earned entrenchment of a term is a function of the number of times it has been projected in the past, or rather, the number of times predicates 'coextensive with it' ([5] p. 95) have been projected in the past. 'Differences of tongue, use of coined abbreviations, and other variations in vocabulary do not prevent accrual of merited entrenchment. Moreover, no entrenchment accrues from the repeated projection of a word except when a word has the same extension each time' ([5] pp. 95f.). Now if Goodman had written that the entrenchment of a term is a function of the number of times a term with the same meaning as it has been projected, we would have an easier task in determining the entrenchment of a term. For we can find out which terms mean the same as other terms by investigating how men use the terms. But Goodman does not believe that terms have the same meaning where this is something distinct from the same extension (that is the two terms in fact applying to the same objects). (See p. 102 note (1).) But on Goodman's view how are we to determine if two terms have the same extension? We cannot without determining the truth or falsity of predictions. For example we can only determine whether the extension of 'grue emerald' and 'emerald' is the same by determining whether future emeralds will be grue. But the whole point of Goodman's rules of projection, which cannot be applied until we have determined entrenchment, was to determine whether it was at all probable that future emeralds will be grue (which is the most we can find out at the moment; we cannot know for certain whether they will be grue). Hence, on Goodman's views about meaning, we cannot even apply Goodman's rules of projection until we have determined by other means what results they will give! This shows the impossibility of determining the relative simplicity of nomological propositions, given Goodman's views about meaning. However, these views are surely mistaken. We did know, for example, before black swans were discovered, that 'swan' and 'white swan' differed in meaning, although we believed that they had the same extension. And whether a term denotes a well-entrenched concept is a matter not merely of its history but of the history of terms having the same meaning as it. I conclude that we must substitute in Goodman's account 'synonymous' for 'coextensive'. This consideration meets difficulties raised in footnote (1) to p. 102.

But, understanding 'projection' in this way, how are we to calculate the number of times a predicate has been projected? A nomological proposition, and so a predicate occurring in it, it will be recalled, is projected if it is adopted when it is supported, unviolated and unexhausted; and by 'adoption' Goodman meant 'something like affirmation as sufficiently more credible than alternative hypotheses' ([5] p. 88). But what constitutes 'affirmation'? Is an hypothesis 'adopted' every time it is used or uttered by any person, or only if it is published in a journal? Are all people on a level – does my uttering an hypothesis constitute just as much a projection of it as Einstein uttering it? Goodman admits to not answering such questions ([5] p. 88), but claims that answers to them would be merely filling out the details of his proposal. But surely without some further filling the proposal is so vague that we are in no position to assess it. Teller ([13] pp. 233f. and pp. 237f.) makes the point that estimating entrenchment 'by simply counting heads seems too crude . . .. Natural phenomena (and our inductive practice is a natural phenomenon) rarely succumb to such blunt attack'. Further, why should the use of a predicate in a past projection entrench it while its frequent use in descriptions does not? Goodman stresses that 'a very familiar predicate may be rather poorly entrenched, since entrenchment depends on frequency of projection rather than upon mere frequency of use' ([5] p. 97). But why should it? Even if no nomological propositions had been formulated about objects of a certain kind always being 'green' the fact that this word had often been used to describe objects seems to make it a natural word to use in hypotheses. Another point is that if a predicate 'A' has been projected n times in so far unfalsified hypotheses it has, if I understand Goodman correctly, the same entrenchment as a predicate 'B' which has been pro-

jected *n* times in an hypothesis subsequently falsified. But surely '*A*'
deserves to be projected in future more than '*B*' does.

In view of these objections, let us try to give a more realistic, if rather
vaguer, picture of the matter. We are born into a community which has
certain ways of describing things; we learn and use those ways. Predicates
in current use are better entrenched than other predicates. However
among predicates in current use the application of some to objects is
explained by scientists in virtue of the application of others to objects.
Colours are explained by wavelength reflection, sounds by air vibrations
etc. This explanation is provided by nomological propositions being
supported, unviolated and unexhausted which correlate predicates of the
latter type with predicates of the former type, e.g. 'all light of such and
such wavelength is green'. Predicates of the 'wavelength' etc. type occur
in various laws, e.g. other laws about the behaviour of light and other
particles. These predicates are much better entrenched than other predicates, because of the wide range of the nomological propositions in which
they occur. Entrenchment is most surely secured by occurrence in nomological propositions currently accepted (in the sense of 'believed') by the
whole scientific community; a smaller degree of entrenchment is provided
by occurrence in nomological propositions currently accepted by part of
the scientific community. It is increased by the eminence (especially in
respect of their success in prediction in related fields) of those who accept
the nomological propositions. Predicates also derive some small entrenchment from their use in nomological propositions subsequently falsified.
If entrenchment in a language is to determine projectibility, the above,
I suggest, gives a more realistic picture of entrenchment than Goodman's
'number of past projections'.

Entrenchment, it will be recalled, was, Goodman claimed, a matter of
both earned and inherited entrenchment. We must now modify the
account which we have given of entrenchment which referred only to
earned entrenchment, by taking account of inherited entrenchment.
Goodman holds that inherited entrenchment has much less weight in
determining total entrenchment than does earned entrenchment. He
introduces the concept of inherited entrenchment by way of the concept of a parent predicate:

A predicate '*P*' is a parent of a given predicate '*Q*' if among the classes
that '*P*' applies to is the extension of '*Q*'; for example the predicate
'army division' is a parent of the predicate 'soldier in the 26th division'.
([5] p. 104).

(The extension of '*Q*' is the class of objects which are *Q*.) And to add examples, 'genus of birds' is a parent predicate of 'duck', 'bagful of marbles' is a parent predicate of 'marble in bag B', and 'colour' is a parent predicate of 'green'. A predicate, unlike a person, will have any number of parents. If of two predicates 'C' and 'D' the best entrenched parent of 'C' is better entrenched than the best entrenched parent of 'D', 'C' has greater inherited entrenchment than 'D'. Hence, Goodman claims, 'marble in bag B' has greater inherited entrenchment than 'marble in zig B', 'applying to marbles in some quite helter-skelter collection'. ([5] p. 104). Hence if neither 'marble in bag B' nor 'marble in zig B' has been projected before, and so neither has any earned entrenchment, 'marble in bag B' will have greater inherited entrenchment, and so greater total entrenchment, and hence, other things being equal, hypotheses using it have greater projectibility than hypotheses using 'marble in zig B'.[1]

This all seems reasonable enough given our previous qualifications on the understanding of 'entrenchment'. Adoption of a system of classification cuts up the world in certain ways. It commits us to certain ways of description, and even if we have not previously used some of them in projection, they still have priority over other ways.

Bibliography

For Goodman's controversy with Carnap and classical statement of the 'new riddle of induction', see:

[1] NELSON GOODMAN 'A Query on Confirmation' *Journal of Philosophy*, 1946, **43**, 383-5.

[2] RUDOLF CARNAP 'On the Application of Inductive Logic' *Philosophy and Phenomenological Research*, 1947, **8**, 133-47.

(1) One detail in Goodman's account which I have so far neglected, because taking it into consideration would not have made any difference to the main points which I have been making about Goodman, is the following. Goodman distinguishes ([5] p. 101 n. 13) entrenchment as antecedent from entrenchment as consequent. When comparing the entrenchment of terms we compare them as one or the other. Earned entrenchment as antecedent depends on the number of past projections in which the term has occurred as antecedent; earned entrenchment as consequent depends on the number of past projections in which the term has occurred as consequent. Inherited entrenchment as antecedent depends on the entrenchment of a parent predicate as antecedent; inherited entrenchment as consequent depends on the entrenchment of a parent predicate as consequent. However, considerations to be put forward in Chapter XI suggest that inherited entrenchment as consequent depends on the entrenchment of what I shall there term an S-parent as consequent.

[3] NELSON GOODMAN 'On Infirmities of Confirmation Theory' *ibid.*, 1947, 8, 147-51.

[4] RUDOLF CARNAP 'Reply to Nelson Goodman' *ibid.*, 1948, 8, 461-2.

[5] NELSON GOODMAN *Fact, Fiction, and Forecast* (first published London, 1955), 2nd edition, Indianapolis, 1965. My references are to the latter edition.

For Goodman's latest amended statement of his rule of projectibility, see:

[6] ROBERT SCHWARTZ, ISRAEL SCHEFFLER and NELSON GOODMAN 'An Improvement in the Theory of Projectibility' *Journal of Philosophy*, 1970, 67, 605-8.

On the attempted solution of the 'grue-green' conflict in terms of a distinction between qualitative and positional predicates, see:

[7] S. F. BARKER and PETER ACHINSTEIN 'On the New Riddle of Induction' *Philosophical Review*, 1960, 69, 511-22; reprinted in P. H. Nidditch (ed.) *The Philosophy of Science,* London, 1968, pp. 149-61.

[8] NELSON GOODMAN 'Positionality and Pictures' *Philosophical Review*, 1960, 69, 523-5; reprinted in P. H. Nidditch (ed.) *op. cit.,* pp. 162-4.

[9] J. S. ULLIAN 'More on "grue" and grue' *Philosophical Review*, 1961, 70, 386-9.

[10] KENNETH SMALL 'Professor Goodman's Riddle' *ibid.*, 1961, 70, 544-52.

[11] SIMON BLACKBURN 'Goodman's Paradox' in Nicholas Rescher (ed.) *Studies in The Philosophy of Science*, Oxford, 1959, pp. 128-42.

On Goodman's understanding of definition, see:

[12] NELSON GOODMAN *The Structure of Appearance,* 2nd edition, Indianapolis, 1966.

For detailed commentary on Goodman's account of entrenchment, see:

[13] PAUL TELLER 'Goodman's Theory of Projection' *British Journal for the Philosophy of Science*, 1969, 20, 219-38.

[14] DAVID H. SANFORD 'Disjunctive Predicates' *American Philosophical Quarterly*, 1970, 7, 162-70.

On the measurement of mathematical simplicity, see:

[15] HAROLD JEFFREYS *Scientific Inference*, Cambridge, 1931, Chapter IV.

[16] JOHN G. KEMENY 'The Use of Simplicity in Induction' *Philosophical Review*, 1953, **62**, 391–408.

[17] MARY HESSE 'Simplicity' in Paul Edwards (ed.) *Encyclopedia of Philosophy*, New York, 1967.

[18] ROBERT ACKERMANN 'A Neglected Proposal Concerning Simplicity' *Philosophy of Science*, 1963, **30**, 228–35.

[19] ROBERT ACKERMANN 'Inductive Simplicity' *ibid.*, 1961, **28**, 152–61. Reprinted in P. H. Nidditch (ed.) *op. cit.*, pp. 124–35.

For a different approach to simplicity, see:

[20] G. SCHLESINGER *Method in the Physical Sciences*, London, 1963, Chapter I.

I

CHAPTER VIII

The Probability of Particular Propositions–I

Chapters VII to XI are concerned to show how quantitative or comparative values are to be given to various hypotheses on various pieces of evidence. In this chapter and the following chapter we shall concern ourselves with the probability of particular propositions on evidence which consists of particular propositions. Often we can say of a particular proposition h on evidence e that its probability is greater or less than that of some other proposition h' on evidence e'. Often too we can go further and assign an exact value to the probability of a particular proposition – we can say that the probability on such and such evidence of the top card being a heart is $1/4$ or the next man I meet in the street being a Roman Catholic is $1/10$. We must now ask how are such probabilities to be calculated on evidence which consists of particular propositions.

Suppose we wish to know the probability of an object a having a property Q – Nijinsky winning the Derby, or Jones living until the age of sixty. We will know about a that it has various properties – A, B, C, D, and so on. Let 'α' denote the conjunction of predicates denoting properties which a is known to possess. (In other words the class 'α' is the intersection of the classes to which a is known to belong.)

One answer to the question of how is the probability of a being Q to be calculated has been provided by Reichenbach ([4] pp. 374f.) for the special case where the sole evidence is that m x's have been observed and $\Phi_m(Q/\alpha) = q$ (that is, mq of them are Q, and $(m - mq)$ of them are $\sim Q$). (See also Salmon [7] pp. 83–96.) He claims that in such a situation we ought to 'posit' q as the value of the probability that a is Q. (I will henceforward neglect the complication that in Reichenbach's view the proposed value of the probability that a is Q on the cited evidence is only 'posited', not necessarily true. I neglect it in view of my arguments in Chapter I for the necessary truth of propositions of epistemic probability.) Adopting the value of $\Phi(Q/\alpha)$ as the probability that a is Q is called following Reichenbach's 'straight rule'. Reichenbach adds the qualification that we can only use this rule if m is quite large, i.e. a considerable number of α's have been examined. Background evidence will show how large the amount of evidence has to be by showing after how many observations predictions in similar fields prove reliable. (A similar account is given by Ayer [6] in attempting to fill out the Frequency Theory of probability.) Reichenbach ([4] p. 375) writes that in a case where we have 'insufficient statistics' we cannot say what the probability is.

But surely if only a few α's have been observed, we can say something about the probability that a is Q, perhaps only that it is high or low. If we have observed six α's and found them all to be Q, that will make the probability of the next α being Q quite high; whereas if we have observed six α's and found them all to be $\sim Q$, that will make the probability of the next α being Q quite low, all this given no further evidence. Another objection to Reichenbach's straight rule is that according to it if a lot of α's have been observed and if all of them are found to be Q, the probability that a is Q is 1, and if all of them are found to be $\sim Q$ the probability that a is Q is 0. Yet it is not in the situation postulated certain that a is Q, nor is it certain that a is $\sim Q$. Reichenbach's rule seems to yield a wrong result. (The intuitive undesirability of this result is used as argument in favour of his own system against Reichenbach's by Carnap – see [5] p. 568.) In any case we require a more general account of how the probability that a is Q is to be calculated, for the evidence is seldom of the narrow kind considered so far.

The axioms of the probability calculus yield a simple theorem which has relevance to this problem, and which yields, I suggest, plausible results. Given a finite number n of hypotheses $h_1 \ldots h_n$, which are exclusive and exhaustive (that is, of logical necessity at least and at most one of them must be true), then for any s and e $P(s/e) = P(s \cdot h_1/e) + P(s \cdot h_2/e) \ldots + P(s \cdot h_n/e)$. (Since '$s$' is logically equivalent to '$(s \cdot h_1)$' v $(s \cdot h_2)$ v $\ldots$ v $(s \cdot h_n)$, when $h_1 \ldots h_n$ are exclusive and exhaustive, the theorem follows from Axiom V and repeated applications of Axiom III.) In that case, by Axiom IV, $P(s/e) = P(s/h_1 \cdot e) P(h_1/e) + P(s/h_2 \cdot e) P(h_2/e) \ldots$ $\ldots + P(s/h_n \cdot e) P(h_n/e)$.
This we can summarize by

$$P(s/e) = \sum_{i=1}^{i=n} P(s/h_i \cdot e) P(h_i/e).$$

$(\sum\limits_{i=1}^{i=n}$ followed by a formula is a short way of writing the sum of the

expression which you obtain by substituting '1' for the '*i*' in the formula, with the expression which you obtain by substituting '2' for the '*i*' in the formula, with the expression which you obtain by substituting '3' for the '*i*' in the formula, and so on, until you reach the expression which you obtain by substituting '*n*' for the '*i*' in the formula.) Let *s* be the proposition that *a* is *Q*, and *e* be our evidence (that *a* is α, and the other evidence, whether the narrow evidence that a certain number of α's have been observed to be *Q* or some wider evidence).

Let us further suppose that there is only a finite number *n* of possible values of $Pr(Q/\alpha), p_1 \ldots p_n$; and let h_1 be the hypothesis that $Pr(Q/\alpha) = p_1$, h_2 be the hypothesis that $Pr(Q/\alpha) = p_2$, and so on. Now we saw earlier (p. 29) that if *e* is merely the evidence that *a* is α, and h_i and *s* are as stated (h_i is '$Pr(Q/\alpha) = p_i$', and *s* is '*a* is *Q*'), then $P(s/h_i. e) = p_i$. Given that, as we have postulated (p. 124), *a* is α is the sum total of our knowledge about *a*, it is hard to see how the addition of any other evidence to *e* can affect the value of $P(s/h_i. e)$. If we know that *a* is α, and that the (complex statistical) probability (that is, roughly speaking, the frequency) of an α being *Q* is $\frac{2}{3}$, then on that evidence the probability of *a* being *Q* must be $\frac{2}{3}$. This must surely remain the probability whatever else we know about the world, so long as we do not know that *a* is different from other α's in some further respect. This, *ex hypothesi*, we do not know. I conclude that, given that *e* includes that *a* is α and no further information about *a*, and h_i and *s* are as stated, $P(s/h_i. e) = p_i$. So our theorem reduces to:

$$P(s/e) = \sum_{i=1}^{i=n} P(h_i/e)\, p_i$$

I will call this formula with the interpretation given above of h_i, p_i, etc. the first form of 'the weighted sum formula'. The probability of *a* being *Q* is given by the sum of possible values of $Pr(Q/\alpha)$, each weighted by the probability of that value being the true value.

However in this form the weighted sum formula is not a great deal of use. It was derived on the supposition that the number of possible values of $Pr(Q/\alpha)$ is finite; and in fact $Pr(Q/\alpha)$ can take any of the infinite number of values between 1 and 0. In that case, as we saw in Chapter V (given that $P(e) \neq 0$) the probability on any evidence of observation *e* that any one of them, apart from 1 or 0, is the true value, will, given our normal mathematics, be zero. If our evidence *e* includes evidence that a proportion q (where $q \neq 1$ or 0) of α's have been observed to be Q, neither 1 nor 0 can be the true value of $Pr(Q/\alpha)$. In that case, given the above formula the probability of s will be 0. But it is clearly false that $P(s/e)$ always equals 0 in such circumstances. Hence the formula does not apply when $Pr(Q/\alpha)$ can take any of an infinite number of values. Nor for this case does it follow from the axioms of the calculus. For, as we saw in Chapter V (p. 72) the probability calculus can have no Axiom III* for one of an infinite number of (exclusive and exhaustive) hypotheses being true; and from this it follows that a theorem corresponding to the theorem on p. 125 cannot be generated for the case where the number of hypotheses h_i is infinite.

Nevertheless, allowing that $Pr(Q/\alpha)$ could have any of the infinite number of values between 1 and 0, Laplace [1] derived the weighted sum formula for this case using an Axiom III*, and proceeded to work with it on the tacit assumption that the probability of any particular value of $Pr(Q/\alpha)$ being the true value was not in general 0. Using his principle of indifference (see my p. 23) he argued that the intrinsic probability of any value of $Pr(Q/\alpha)$ being the true value was the same. From this it follows by Bayes's Theorem that the posterior probability that any value p of $Pr(Q/\alpha)$ was the true value is proportional to its accuracy, that is to the probability of the observed data given that $Pr(Q/\alpha) = p$. From that Laplace derived his famous rule of succession ([1] p. 433f.). This states, in our terminology, on evidence that m α's are observed and that the proportion of them found to be Q is q $(\Phi_m(Q/\alpha) = q)$, the probability that a is Q is

$$\frac{mq + 1}{m + 2}$$

However, interesting results in this field including (with qualifications) Laplace's rule can be derived from the axioms of the probability calculus without the use of Laplace's dubious assumptions. We saw on p. 125 that it is a theorem of the calculus that given n exclusive and exhaustive hypotheses $h_i \ldots h_n$, $P(s/e) = \sum_{i=1}^{i=n} P(s/h_i \cdot e) \, P(h_i/e)$. Now let each of the h_i state, not that $Pr(Q/\alpha)$ has some precise value but that it lies within some interval. Let the continuum between 1 and 0 be divided into n equal intervals. Let h_1 state that $Pr(Q/\alpha)$ lies between 0 and $1/n$; h_2 that $Pr(Q/\alpha)$ lies between $1/n$ and $2/n \ldots$ and h_n that $Pr(Q/\alpha)$ lies between $(n-1)/n$ and 1. Let us make this more precise as follows. Let p_i be i/n (for example p_1 will be $1/n$ and p_n be 1). Let h_i be the hypothesis that $p_{i-1} < Pr(Q/\alpha) \leqslant p_i$

(except that h_1 be the hypothesis that $0 \leqslant Pr(Q/\alpha) \leqslant p_i$). Let n be a very large finite number. By arguments similar to those used before we may conclude that for any e $P(s/h_i. e)$ has the same value as where e is merely that a is α. Intuitively, as we take n larger, and so h_i makes a narrower claim about the range within which $Pr(Q/\alpha)$ lies (that it lies between p_i and some value slightly less than p_i), $P(s/h_i. e)$ gets closer to p_i. We saw earlier that the probability that a is Q on evidence that a is α and $Pr(Q/\alpha)$ $= p_i$ is p_i. Intuitively if h_i says, not that $Pr(Q/\alpha)$ is p_i but that it is p_i or very slightly less, $P(s/h_i. e)$ remains to a high degree of approximation p_i. So we reach the result that for very large n, to a degree of approximation which we can ignore, $P(s/e) = \sum\limits_{i=1}^{i=n} P(h_i/e) \, p_i$. I will call this formula with the new interpretations of p_i, h_i, etc. given above the second form of the weighted sum formula.[1]

Intuitively in either version the formula is plausible. The epistemic probability of a being Q is greater, the greater the statistical probability of objects like a being Q. We do not know what is the statistical probability of objects like a being Q, yet perhaps we can to some extent calculate the epistemic probability that the statistical probability of objects like a being Q has different values or lies within different intervals. Surely the probability that a is Q is greater, the greater the epistemic probability of theories ascribing a high value to the statistical probability that objects like a are Q. Our formula brings out this requirement. Talk about the probability of a probability does not put us on the slippery slope of an infinite regress, for we are talking about two different kinds of proba-

---

(1) This can be expressed in terms of the well-known probability distribution function $F(p)$ as follows. Let $F(p)$ state the probability given e that $Pr(Q/\alpha)$ has a value between 0 and p inclusive. Then by this second form of the weighted sum formula

$$P(s/e) = \sum_{i=1}^{i=n} \left\{ F\left(\frac{i}{n}\right) - F\left(\frac{i-1}{n}\right) \right\} \frac{i}{n}$$

If we let n tend to infinity, we can rephrase this formula in terms of the probability density function $f(p)$, where $f(p)$ is the probability density for a value p of $Pr(Q/\alpha)$, as follows:

$P(s/e) = \int_0^1 f(p) \, p \, dp$

(on the assumption that $F(p)$ has a derivative at all points p, $0 < p < 1$). The probability density $f(p)$ is defined as $\dfrac{d}{dp} F(p)$. The probability density function of $Pr(Q/\alpha)$ measures, roughly speaking, the relative probability of getting $Pr(Q/\alpha)$ very close to different values. If the density function has a high value for $p = \frac{1}{2}$, and a low value for $p = \frac{1}{10}$, then it is much more probable that $Pr(Q/\alpha)$ is close to $\frac{1}{2}$ than close to $\frac{1}{10}$.

bility. We are talking about the epistemic probability, the probability on evidence, that a certain value of a complex statistical probability holds. Whether a certain value of the statistical probability does hold is an empirical matter; how much the evidence renders probable the claim that it does is, for given evidence, a logical matter.

The second form of the weighted sum formula is however the only one which can be used when $Pr(Q/\alpha)$ can take any of the infinite number of values between 0 and 1. One very simple example of its application is the following. Suppose that $Pr(Q/\alpha)$ is equally likely to lie within any interval of specified length, however small. There are n such intervals. Then for every i $P(h_i/e) = 1/n$. So

$$P(s/e) = \frac{1}{n} \sum_{i=1}^{i=n} p_i = \frac{1}{n} \sum_{i=1}^{i=n} \left(\frac{i}{n}\right) = \frac{1}{n} \left\{ \frac{n(n+1)}{2n} \right\} = \frac{n+1}{2n}.$$

This for very large n is, to a degree of approximation which we can ignore, $\frac{1}{2}$. This result seems intuitively correct. If $Pr(Q/\alpha)$ is equally likely to lie within any interval of given length between 0 and 1, however small, then the probability that a is Q is $\frac{1}{2}$. Henceforward by 'the weighted sum formula' I shall refer to the second form of the formula.

Now by Bayes's Theorem $P(h_i/e)$ in this formula will, given that e is our present evidence and so $P(e) \neq 0$, be equal to $\dfrac{P(h_i)\,P(e/h_i)}{P(e)}$. If we suppose, not, as Laplace supposed, that each value of $Pr(Q/\alpha)$ was equally likely to be the true value, but rather that each hypothesis h_i (that $Pr(Q/\alpha)$ is i/n or very slightly less) is equally likely to be true, we can now derive Laplace's rule of succession. For in this case, $P(h_i)$ and $P(e)$ being the same for each h_i, $P(h_i/e)$ will be proportional to $P(e/h_i)$. h_i states that $Pr(Q/\alpha)$ is i/n or very slightly less, and so to an approximation which we can ignore for large n, $P(e/h_i)$ is $P(e/Pr(Q/\alpha) = i/n)$. Now let our evidence e be that m α's are observed and mq of them found to be Q (that is, a proportion q of them are Q). If we assume that m α's being observed is equally likely on any hypothesis h_i (an assumption which we shall need to investigate in Chapter X) $P(e/h_i)$ and so $P(h_i/e)$ will be proportional to $P(\Phi_m(Q/\alpha) = q/Pr(Q/\alpha) = i/n$. m α's are observed). This by the result of p. 29 is

$$\frac{m!}{(m-mq)!mq!} \left(\frac{i}{n}\right)^{mq} \left(\frac{n-i}{n}\right)^{m-mq}.$$

Now given that $P(h_i/e)$ is proportional to this figure, it can easily be proved that $P(s/e)$, which by the weighted sum formula is $\sum_{i=1}^{i=n} P(h_i/e)\, p_i$ is $\dfrac{mq+1}{m+2}$, which is what Laplace's rule of succession states. (Outlines of a proof using an integral instead of a sum may be found in Keynes [2] pp. 375f.)

The rule of succession states that, in the absence of other evidence, if m α's have been observed and mq of them found to be Q, then the probability of the next α, that is a, being Q is $\dfrac{mq + 1}{m + 2}$. If we know that 21 out of 43 α's are Q, the probability of the next one being Q is $\dfrac{21 + 1}{43 + 2} = \dfrac{22}{45}$.

If no α's have been observed, the probability of the next one being Q is $\frac{1}{2}$. If all of 10 α's observed have been found to be Q, the probability of the next one being Q is $\frac{11}{12}$. The rule was derived by me on the supposition that the intrinsic probability of $Pr(Q/\alpha)$ lying within any interval however small was the same as that of it lying within any other interval of the same size (but by Laplace on the assumption that the intrinsic probability of $Pr(Q/\alpha)$ having any one value was the same as that of it having any other value).

Various objections have been made to Laplace's rule. (See e.g. Keynes [2] pp. 371–83, and Kneale [3] pp. 201–7.) Many concern the method of its derivation which I have criticized above. Others arise from a confusion on the part of critics or expositors of the rule between the epistemic probability that a is Q, the statistical probability $Pr(Q/\alpha)$, and the epistemic probability on different pieces of evidence that the statistical probability $Pr(Q/\alpha)$ has a certain value. I claim to have phrased my account so as to be immune to all such objections. The most telling objection, to my mind, is the objection that the rule leads to inconsistent ascriptions of probability. Let 'Q', 'R', and 'S' denote mutually incompatible properties such that any α must have one or other of them. Suppose that of n α's observed np_1 were Q, np_2 were R, and np_3 were S. $p_1 + p_2 + p_3 = 1$. We would conclude by Laplace's rule that the probability of a being Q is $\dfrac{np_1 + 1}{n + 2}$, of a being R is $\dfrac{np_2 + 1}{n + 2}$, of a being S is $\dfrac{np_3 + 1}{n + 2}$. Yet $\dfrac{np_1 + 1}{n + 2} + \dfrac{np_2 + 1}{n + 2} + \dfrac{np_3 + 1}{n + 2} > 1$. But since a must be either Q, or R, or S, but can have no more than one of these properties, then by Axiom III the value of this sum ought to be 1. (See Kneale [3] p. 204 for this objection.)

Now what this objection shows is that Laplace's rule cannot hold for each of three incompatible properties 'Q', 'R', and 'S', one of which must belong to any α. However before we can derive Laplace's rule we have to make some supposition about intrinsic probability. I replaced Laplace's supposition that each value of $Pr(Q/\alpha)$ had the same intrinsic probability by the supposition that the intrinsic probability that $Pr(Q/\alpha)$ lies within any interval, however small, is the same. Only on such a supposition can the rule be applied. But the assumption (where 'Q', 'R' and 'S' denote incompatible properties, one of which must belong to any α) that both the intrinsic probability of $Pr(Q/\alpha)$ lying within any two intervals of the same length is the same, and the intrinsic probability of $Pr(R/\alpha)$ lying within any two intervals of the same length is the same, is inconsistent with the assumption that the intrinsic probability of $Pr(S/\alpha)$ lying within any two intervals of the same length is the same. This is easy to see. The claim that the intrinsic probability of $Pr(Q/\alpha)$ lying within any two intervals of the same length is the same entails the claim that the intrinsic probability of $Pr(\sim Q/\alpha)$ lying within any two intervals of the same length is the same. We supposed the range of possible values of probabilities between 1 and 0 to be divided into n equal intervals of length $1/n$. Then the intrinsic probability that $Pr(\sim Q/\alpha)$ lies within any one such interval will be $1/n$. $Pr(\sim Q/\alpha) = Pr(R/\alpha) + Pr(S/\alpha)$ (since an α which is $\sim Q$ must be R or S). Then the intrinsic probability that $\{Pr(R/\alpha) + Pr(S/\alpha)\}$ lies within any interval of length $1/n$ will be $1/n$. But the intrinsic probability of $Pr(R/\alpha)$ lying within any two intervals of equal length is said to be the same. In that case the intrinsic probability that $Pr(R/\alpha)$ lies within any interval of length $1/n$ will be $1/n$. The same applies to $Pr(S/\alpha)$. Now consider the interval I which includes 0. $Pr(R/\alpha) + Pr(S/\alpha)$ can lie in I only if both $Pr(R/\alpha)$ and $Pr(S/\alpha)$ lie in I. Consider the interval I' next to I. $Pr(R/\alpha) + Pr(S/\alpha)$ can lie in I' only if either $Pr(R/\alpha)$ or $Pr(S/\alpha)$ lies in I. Hence if the probability that $\{Pr(R/\alpha) + Pr(S/\alpha)\}$ lies in I is $1/n$ and the probability that it lies in I' is $1/n$, then either the probability that $Pr(R/\alpha)$ lies in I or the probability that $Pr(S/\alpha)$ lies in I must exceed $1/n$, which contradicts our original supposition. I conclude that we cannot make the supposition for each of three incompatible properties z that the intrinsic probability that $Pr(z/\alpha)$ lies within any interval of the same size, however small, is the same. So the earlier demonstration that application of Laplace's rule leads to inconsistent ascriptions of probability does not show that Laplace's rule is at fault, but that the assumption about epiprobability necessary for its application cannot be made. I conclude that Laplace's rule only applies given an assumption about intrinsic probability which cannot always be made.

The question therefore arises as to what governs $P(h_i)$, what determines the distribution of the intrinsic probabilities of $Pr(Q/\alpha)$ lying within different intervals. One consideration is that adduced in the last chapter that (if 'α' consists only of projectible predicates) $\Pi(Q/\alpha)$ and so $Pr(Q/\alpha)$ is more likely to be 1 or 0 than to have any other value, and so more likely to lie within an interval which includes 1 or 0 than in any other interval of equal length. Another consideration is the consideration of width. $Pr(Q/\alpha)$ is intrinsically more likely to lie within high intervals (intervals including values close to 1), the wider is the predicate 'Q'. Thus suppose we are considering what will be the colour of a certain house a. Let 'Q_1' be 'of any colour except black', and 'Q_2' be 'magenta'. My suggestion is that $Pr(Q_1/\alpha)$ is intrinsically more likely to be found in intervals which include high values in the probability continuum, than is $Pr(Q_2/\alpha)$. That this is so can be seen from the weighted sum formula. Let s_1 be 'a is Q_1' and s_2 be 'a is Q_2'. Since s_2 entails s_1 and $P(\sim s_2 . s_1) \neq 0$, by Principle E (p. 51) $P(s_2) < P(s_1)$. If we put e as a tautology, then the weighted sum formula (for very large n) yields the following result: $P(s_1) = \sum\limits_{i=1}^{i=n} P(h_i^1)p_i^1$,

where the h_i^1 are exclusive and exhaustive hypotheses about the intervals within which $Pr(Q_1/\alpha)$ lies, and the p_i^1 are the upper bounds of such intervals. Similarly the weighted sum formula yields the result that $P(s_2) = \sum\limits_{i=1}^{i=n} P(h_i^2)p_i^2$ where the h_i^2 are exclusive and exhaustive hypotheses about the intervals within which $Pr(Q_2/\alpha)$ lies, and the p_i^2 are the upper bounds of such intervals. It can be seen that for $P(s_1)$ to be greater than $P(s_2)$, the $P(h_i^1)$ must by and large have larger values for higher values of $Pr(Q_1/\alpha)$ than do the $P(h_i^2)$ for higher values of $Pr(Q_2/\alpha)$. So the wider a predicate 'Q' is, the greater the intrinsic probability of $Pr(Q/\alpha)$ being found in intervals which include high values of the probability continuum from 0 to 1.

So we have seen that $P(h_i)$, where h_i is a hypothesis that $Pr(Q/\alpha)$ lies within some narrow interval, is higher if the interval includes 1 or 0 (where 'α' consists only of projectible predicates), and is higher by and large for higher intervals in the probability continuum the wider is 'Q'. However it does not seem possible to lay down any more precise rules for determining $P(h_i)$. Nor is it necessary where there is a substantial amount of evidence of a kind which makes theories h_i differ from each other markedly in their accuracy, for example where our sole evidence is that many α's have been observed and a certain proportion of them found to be Q. For it is easy to see that in that case $P(h_i/e)$ are affected much less by $P(h_i)$ than by $P(e/h_i)$, the other factor which determines the relative values of $P(h_i/e)$. (We saw earlier that $P(h_i/e) = \dfrac{P(e/h_i)P(h_i)}{P(e)}$. $P(e)$ is a constant for variable h_i and so may be ignored.) For example in any field where the evidence is that 20 α's have been observed and 18 of them found to be Q, it would be agreed that to a high degree of approximation the probability of a being Q is $\frac{18}{20} = 0\cdot9$. Yet this result will only come out of the weighted sum formula if we suppose that $P(h_i)$ is to all intents and purposes constant for different h_i.[1] This means, for example, that we can ignore the width of the predicate 'Q'. I shall therefore henceforward assume that, where h_i is a hypothesis about which of a large number of equal narrow intervals in the probability continuum between 1 and 0 is occupied by $Pr(Q/\alpha)$, $P(h_i/e)$ is directly proportional to $P(e/h_i)$. That is to say a hypothesis that $Pr(Q/\alpha)$ lies within a very narrow interval which includes p is, to a degree of approximation which we will henceforward ignore, proportional to the accuracy of '$Pr(Q/\alpha) = p$'.

So far in this chapter I have been concerned to expound a form of the weighted sum formula, and to develop and justify its consequences for the case where our sole evidence is that a number of α's have been observed and a certain proportion of them found to be Q. The formula, to repeat it, is that, when we take n very large, to a degree of approximation which we can ignore,

$$P(s/e) = \sum_{i=1}^{i=n} P(h_i/e)\, p_i.$$

(In this formula s is the proposition that a is Q, all we know about a is that it has the conjunction of properties α, the h_i are exclusive and exhaustive hypotheses about which of n equal intervals is occupied by $Pr(Q/\alpha)$, and the p_i are the upper bounds of those intervals.) I shall conclude this chapter by stating and justifying some wider consequences of this formula, and in the next chapter show in more detail how the probability that a is Q is to be calculated on wider evidence than that considered so far.

One consequence of the weighted sum formula is that the discovery that b, an already known α, is Q, always confirms s, the proposition that a is Q. The new piece of evidence which we will call e' is more probable

(1) If we suppose that $P(h_i)$ is constant then for different h_i $P(h_i/e)$ is proportional to $P(e/h_i)$. If we further suppose, as is natural, that 20 α's are equally likely to be observed on any hypothesis h_i, then $P(e/h_i)$ (and so $P(h_i/e)$) will be proportional to $P(\Phi_{20}(Q/\alpha) = 0\cdot9/Pr(Q/\alpha) = p_i$. 20 α's are observed). Then by the formula given on p. 29 for $p_i = 0\cdot9$, $P(h_i/e)$ is proportional to $0\cdot2846$. Whereas for $p_i = 0\cdot5$, $P(h_i/e)$ is proportional to $0\cdot0001$. Similar low values result for other values of p_i not close to $0\cdot9$. Hence it can be seen that it is far more probable that $Pr(Q/\alpha)$ lies within a narrow range including $0\cdot9$ than that it lies outside this range. However this result will not in general be obtained if $P(h_i)$ is significantly different for different h_i.

on hypotheses h_i which state that $Pr(Q/\alpha)$ has a high value than on hypotheses h_i which state that it has a low value, and so by Principle B, e' confirms the former hypotheses more than it does the latter ones (in the sense stated on p. 4). Hence since the probability that one of the n hypotheses is true is the same as before (viz. 1), e' increases the probability of hypotheses h_i which state that $Pr(Q/\alpha)$ has a high value and decreases the probability of hypotheses h_i which state that $Pr(Q/\alpha)$ has a low value. Hence, by the weighted sum formula, it increases the probability of, that is confirms, s.

A second consequence of the weighted sum formula is this. If $Pr(Q/\alpha)$ is very likely to be close to one of two values, say 0·01 and 0·02, and very unlikely to have any other value, then the probability that a is Q is approximately midway between these two values, in the case cited 0·015. This result is surely correct. Suppose some physical theory makes it very likely that the (complex statistical) probability of an atom of a certain kind disintegrating within sixty seconds is either approximately 0·01 or approximately 0·02, neither value being more likely than the other, then on the evidence of that theory, the probability of the next atom of this kind disintegrating within sixty seconds would surely be approximately 0·015 (even though this is almost certainly not the probability of *an* atom of this kind disintegrating within sixty seconds). After all, the sensible betting odds on the proposition that the next atom will disintegrate within sixty seconds would be about 0·015 to 0·985.

A third consequence of the weighted sum formula is that on evidence of observation e which reports that various objects, apart from a, were observed and found to have various properties, the probability that a is Q will never be 1 or 0. We saw earlier that it was an objection to Reichenbach's 'straight rule' that in such a case it sometimes yielded the result that the probability was 1 or 0. For even if the evidence makes it far more likely that $Pr(Q/\alpha)$ lies within a very small interval which includes 1 than that it has any other value, other hypotheses h_i about the interval within which $Pr(Q/\alpha)$ lies will still have some probability on e greater than 0. This is because for each such hypothesis h_i, as we have seen, $P(h_i/e) = \dfrac{P(e/h_i)\,P(h_i)}{P(e)}$. $P(e)$, as we have seen, is greater than zero. For each h_i, $P(h_i) > 0$; since, intuitively, any such hypothesis can be confirmed by some evidence e, it must by considerations adduced on pp. 48f. have intrinsic probability greater than 0. $P(e/h_i)$ only equals 0, since e is evidence of observation, if e is incompatible with h_i. But no reports of observations can be incompatible with any h_i. An h_i is a claim about what would be within an interval the proportion of α's which are Q in a population in which what was physically most probable occurred (see pp. 18f.), and that claim is compatible with any claim about the actual proportion of α's observed to be Q or any other observational reports about the actual world (although of course some h_i are very improbable given some e). Since $P(e/h_i)$, $P(h_i)$, and $P(e)$ are all greater than 0 for all h_i, so must be $P(h_i/e)$. Since theories incompatible with any given h_i have probabilities on the cited evidence greater than 0, by Axiom III no one theory h_i will have a probability of 1 on that evidence. But by the weighted sum formula the probability that a is Q can only be 1 if the probability that $Pr(Q/\alpha)$ lies in any interval not including 1 is 0; and the probability that a is Q can only be 0 if the probability that $Pr(Q/\alpha)$ lies in any interval not including 0 is 0.

A fourth and very important consequence of the weighted sum formula is this. If it is much more likely that $Pr(Q/\alpha)$ lies within a certain range, large or small (e.g. between 0·1 and 0·3) than that it has any value outside that range, then the probability that a is Q will be within or very close to that range. Hence if it is much more likely that $Pr(Q/\alpha)$ lies within a small interval which includes a value q than that it has any value outside that interval, then the probability that a is Q is approximately q. This situation arises, we have seen, where a large number of α's have been observed and $\Phi(Q/\alpha) = q$. For in that case theories h_i ascribing to $Pr(Q/\alpha)$ a value significantly different from q will have proved highly inaccurate and so, we saw, for such theories $P(h_i/e)$ is very low, and hence the disjunction of theories ascribing to $Pr(Q/\alpha)$ a value close to q will be far more probable than the disjunction of all theories ascribing to $Pr(Q/\alpha)$ any other value. Hence our formula gives the same results as Reichenbach's rule to a high degree of approximation for those cases where, according to Reichenbach, the rule applies. (But note that if $\Phi(Q/\alpha) = 1$ or 0, according to Reichenbach the probability of a being Q is respectively 1 or 0. On our formula it is very close to 1 or 0, as the case may be, not exactly 1 or 0.) Our formula however has, as we have seen, wider application than Reichenbach's rule.

Bibliography

[1] P. S. DE LAPLACE *Théorie Analytique des Probabilities* in *Œuvres*, Paris, 1847, Vol. 7, Book 2, Chapter 6.

[2] J. M. KEYNES *A Treatise on Probability,* London, 1921, Chapter 30.

[3] W. C. KNEALE *Probability and Induction,* Oxford, 1949, Part III.

[4] H. REICHENBACH *The Theory of Probability,* Berkeley and Los Angeles, 1949, Chapter 9.

[5] RUDOLF CARNAP *Logical Foundations of Probability,* 2nd edition, London, 1962, pp. 567-9.

[6] A. J. AYER 'On the Probability of Particular Events' in *The Concept of a Person,* London, 1963.

[7] WESLEY C. SALMON *The Foundations of Scientific Inference,* Pittsburgh, 1967, pp. 83-96.

CHAPTER IX

The Probability of Particular Propositions–II

In Chapter VIII we reached the conclusion that (to a degree of approximation which we could ignore) the probability that an object a had a property 'Q' was for large n equal to $\sum_{i=1}^{i=n} P(h_i/e)p_i$, where $p_i = i/n$, h_i is the probability that $p_{i+1} > Pr(Q/\alpha) \geqslant p_i$, α is the conjunction of predicates which according to e a possesses, and e is our total available evidence. It was a consequence of this that if it is far more likely that $Pr(Q/\alpha)$ lies within a small interval including q than that it has any other value, then the probability that a is Q is q (to a degree of approximation which we can ignore). This situation will arise when a large number m of α's are observed and $\Phi_m(Q/\alpha) = q$. Hence we can conclude with Reichenbach that when our sole evidence is that a large number m of α's have been observed and $\Phi_m(Q/\alpha) = q$, the probability of a being Q is q.

In practice however we are not very likely to have as evidence a value of $\Phi_m(Q/\alpha)$, for a large number m of α's observed. One common situation is where we have information about proportions of Q's in wider classes than 'α', and another less common but philosophically interesting situation is where we have information about proportions of Q's in narrower classes than 'α'. We will deal mainly with the former case, but in conclusion apply our results very briefly to the latter case. The former case is exemplified by a typical problem confronting the insurance company.

Suppose that we are attempting to calculate the probability of a certain man, Jones (a) dying before sixty (Q). We know many of the properties of Jones. He is currently aged thirty-four (A), smokes ten cigarettes a day (B), had a parent who suffered from asthma (C), has a surname beginning with J (D) etc. We will hardly ever have any value for $\Phi(Q/\alpha)$ and if we do it will usually be based on very few instances (we will have observed few α's), and so will not be of great weight against other evidence in calculating the probability of $Pr(Q/\alpha)$ lying within some interval. On the other hand, where A, B, C, D, etc. are properties which belong to the conjunction α, i.e. properties known to be possessed by a, we will often have as evidence values of $\Phi(Q/A)$, $\Phi(Q/C)$, $\Phi(Q/B \cap D)^{(1)}$ and so on. We will have information about the proportion of men aged thirty-four who die before sixty, about the proportion of those whose parents have asthma who die before sixty, about the proportion of those whose surname begins with J and smoke ten cigarettes a day who die before sixty and so on. The first problem to be solved in this chapter is how is the probability of a being Q to be calculated on evidence that e is an α and on evidence of the values of $\Phi(Q/A)$, $\Phi(Q/B)$, $\Phi(Q/B \cap C)$, $\Phi(Q/D)$ etc. where a number of A's, B's, $(B \cap C)$'s, D's etc. have been observed (which I shall call the statistical data), where A, B, C, D etc. are properties which are known to belong to the conjunction α.

The solution which has been proposed by both Reichenbach [1] and Ayer [4] is as follows. Reichenbach recommends that we consider 'the narrowest class for which reliable statistics can be compiled' ([1] p. 374), and Ayer, working provisionally within the confines of a Frequency Theor of probability similar to Reichenbach's, suggests that we seek 'the narrowest class in which the property occurs with extrapolatable frequency' ([4] p. 202).[(2)] They suppose that for each x there is some value of m,

(1) Recall that $(B \cap D)$ is the class of objects which are both B and D. Hence $\Phi(Q/B \cap D)$ is the proportion of Q's among objects observed to be both B and D.

(2) Salmon prefers to say that 'the single case should be referred to the *broadest homogeneous reference class* of which it is a member' ([5] p. 91, his italics) but his proposal amounts to the same as Reichenbach's. For given that a is A, B, C, D etc. and we are investigating the probability of its being Q, a class A is said to be homogeneous if and only if there is no class x such that $Pr(Q/A \cap x) \neq Pr(Q/A)$. However, Salmon admits, we have in general to be satisfied with classes which are 'epistemically homogeneous' ([5] p. 92). A class A is said to be 'epistemically homogeneous' if and only if there is no class x *known* such that $Pr(Q/A \cap x) \neq Pr(Q/A)$. Salmon supposes that 'a probability is something that has to be established inductively', that is, we infer $Pr(Q/A)$ by the 'straight rule' from $\Phi(Q/A)$, given that there are enough observations. In that case we infer that the class A is homogeneous if and only if there is no class x such that $\Phi(Q/A \cap x) \neq \Phi(Q/A)$, where the number of observations is sufficiently great. Then suppose that A is the broadest homogeneous reference class, because if for any x $\Phi(Q/A \cap x) \neq \Phi(Q/A)$, $\Phi(Q/A \cap x)$ is not based on sufficient observations to allow inference to $Pr(Q/A \cap x)$. Then either A is the 'narrowest reference class for which reliable statistics can be compiled' or there is a narrower one $(A \cap y)$. In latter case $\Phi(Q/A \cap y) = \Phi(Q/A)$, for if they were not equal A would not be the broadest homogeneous reference class. So the results given by the two methods are the same.

such that when m or more x's have been observed, $\Phi(Q/x)$ gives us not merely the best available, but a 'reliable' estimate of $Pr(Q/x)$. What we should do, they suggest, is to find that conjunction γ of predicates from among those which make up α which denotes the narrowest class in which $\Phi(Q/\gamma)$ does give a 'reliable' estimate of $Pr(Q/\gamma)$. (A class U is narrower than a class W if and only if $U \subset W$, that is the class U is included in the class W but the class W is not included in the class U; if a relationship of this kind does not hold between two classes, they are not comparable in respect of narrowness.) Suppose 'Q' is 'dies before sixty', 'A' is 'heavy smoker', 'B' is 'heavy drinker', 'C' is 'married', and that we know that Jones is married and is a heavy smoker and heavy drinker. Suppose too that we have as evidence that A's, $(A \cap B)$'s, and $(A \cap B \cap C)$'s were observed, that $\Phi(Q/A)$, $\Phi(Q/A \cap B)$, and $\Phi(Q/A \cap B \cap C)$ had certain specified values, and that $\Phi(Q/A \cap B)$ was derived from a large enough sample to give us a 'reliable' estimate of $Pr(Q/A \cap B)$, but that $\Phi(Q/A \cap B \cap C)$ was not derived from a large enough sample to give us a 'reliable' estimate of $Pr(Q/A \cap B \cap C)$. Then on this solution the probability on the evidence of Jones dying before sixty is given by the value of $\Phi(Q/A \cap B)$.

One difficulty with this solution concerns what is meant by saying that an estimate is 'reliable' and what grounds we have for taking one lot of statistical data as 'reliable' and not another lot. But I pass over this in order to concentrate on what seems to me a more radical difficulty. This is that there may be no one 'narrowest class in which the property occurs with extrapolatable frequency', but two or more classes, none of them narrower than any other, which have extrapolatable frequencies, while no narrower class has an extrapolatable frequency, and which lead to different values of the probability that a is Q. Thus a may be known to be A, B, C and D. We may have as evidence the values of $\Phi(Q/A \cap B)$, $\Phi(Q/A \cap C)$, $\Phi(Q/A \cap B \cap C)$, $\Phi(Q/A \cap C \cap D)$, and $\Phi(Q/A \cap B \cap C \cap D)$. $\Phi(Q/A \cap B)$ may be extrapolatable, while $\Phi(Q/A \cap B \cap C)$, $\Phi(Q/A \cap B \cap D)$. and so *a fortiori* $\Phi(Q/A \cap B \cap C \cap D)$ be not extrapolatable; $\Phi(Q/A \cap C)$ may also be extrapolatable while $\Phi(Q/A \cap C \cap D)$ is not extrapolatable. Then $(A \cap B)$ and $(A \cap C)$ are both classes in which Q occurs with extra-polatable frequency, while no narrower class is such a class. Yet neither $(A \cap B)$ nor $(A \cap C)$ is a class narrower than the other and $Pr(Q/A \cap C)$, obtained by extrapolation from $\Phi(Q/A \cap C)$, may not equal $Pr(Q/A \cap B)$ obtained by extrapolation from $\Phi(Q/A \cap B)$. Despite their talk of '*the* narrowest class', Reichenbach and Ayer both admit that this may happen. But they both seem to claim that in these circumstances there are no rational considerations which would allow us to calculate the probability that a is Q. For Reichenbach, 'the calculus of probability cannot help in such a case . . .. The logician can only indicate a method by which our knowledge may be improved' – which is to look for more evidence ([1] p. 375). For Ayer, when generalizations are conflicting, a bettor may reach one estimate of probability or another, but 'there is no objective standard of correctness by which this could be measured' ([4] pp. 207f). But, contrary to Reichenbach and Ayer, there are rational considerations in such cases. Suppose our information is that a is A, B, C and D, that $\Phi(Q/A) = 0 \cdot 015$, that $\Phi(Q/A \cap B) = 0 \cdot 01$, that $\Phi(Q/A \cap C) = 0 \cdot 02$ and that these are extrapolatable frequencies, we can surely conclude at least this about the probability that a is Q; that it lies somewhere in the region of $0 \cdot 01$ and $0 \cdot 02$; that it would be irrational for a man to conduct his betting on any other supposition.

In the Logical Theory tradition Carnap's account in *Logical Foundation: of Probability* and *The Continuum of Inductive Methods* of how individual probabilities are to be calculated from evidence about proportions in observed samples is over-simple in similar ways, though Carnap would agree that his account in these works was over-simple. Carnap assumes that there is some one true description of a, say as an object possessing a conjunction of properties γ, and that the probability of a being Q is to be calculated from $\Phi(Q/\gamma)$, the size of population from which the observed sample is selected, and a certain prior probability.[1] But, as Achinstein [6] and Hesse [7] have pointed out, the extent and variety of the similarities between a and other objects in respect of all properties other than Q, and the proportion of objects similar in various ways to a which are Q, is relevant to the probability that a is Q; the probability that a is Q is not determined merely by what proportion of other objects have one particular conjunction of properties in common with a. Thus if a has many properties δ in common with one group of objects all of which are Q, and many properties ϵ in common with another group of objects all of which are Q, when δ and ϵ are different conjunctions of properties, these two facts raise considerably the probability that a is Q.

(1) See, for example, Carnap [2] p. 15. 'For any i, $c\,(h_i, e_Q) = G(k, s, s_i)$.' This states that the degree of confirmation of h_i by e_Q is a function of and only of k, s, and s_i. h_i is the hypothesis that some individual as yet unobserved has the property Q_i. e_Q states of each member of a population which of the Q-properties, Q_1, Q_2 and so on it has. (When you have stated which Q-property an individual has, you have exhaustively described its characteristics.) s is the number of individuals in the population, s_i is the number which are Q, and k is 'a logical, i.e. non-empirical, magnitude'.

So we must develop a more satisfactory account of this matter. I begin with the results of Chapter VIII that the probability on evidence e that a is Q is for large n equal to $\sum_{i=1}^{i=n} P(h_i/e)p_i$, with the interpretation of the terms as repeated at the beginning of this chapter. We can calculate the quantitative probability on evidence e that a is Q, if we can calculate the quantitative probabilities of different hypotheses of the form '$0.006 > Pr(Q/\alpha) \geqslant 0.007$', '$0.007 > Pr(Q/\alpha) \geqslant 0.008$', and so on. In general this will not be possible, since we do not in general have precise enough criteria to enable us to give to different such hypotheses quantitative probabilities. We saw in Chapter VIII that in so far as $P(h_i)$ was a significant factor in the determination of $P(h_i/e)$ (because $P(e/h_i)$ did not differ greatly for different h_i), it was not often possible to determine an exact quantitative value for $P(h_i/e)$; one could only say that it lay within some rough range, was greater or less than some other value. However sometimes, as we shall see, our criteria do allow us to give sufficiently precise values to the different $P(h_i/e)$ to enable us to calculate a fairly precise value for the probability on e that a is Q; this will happen if $P(e/h_i)$ or $P(h_i)$ differs sharply for different h_i. To the extent to which any one theory, which we will call h_q, of the form '$p_{i-1} > Pr(Q/\alpha) \geqslant p_i$' is made more probable by the evidence than all the others taken together, to that extent the probability that a is Q comes closer to the value of p_i which occurs in h_q and which we will call q. For the greater the probability of h_q, the closer will $\sum_{i=1}^{i=n} P(h_i/e)p_i$ come to q. If it is much more probable that $Pr(Q/\alpha)$ lies within a certain small interval including q than that it has any other value, then the probability that a is Q will be approximately q. But if the evidence only makes it *fairly* probable that $Pr(Q/\alpha)$ is close to any particular value, but very probable that it lies within a certain range, then all we can say is that the probability that a is Q lies within or very close to that range. I shall now show how evidence that various A's, B's, $(C \cap D)$'s and so on were observed and that $\Phi(Q/A)$, $\Phi(Q/B)$, $\Phi(Q/C \cap D)$ and so on had certain values can give to different theories about the interval within which $Pr(Q/\alpha)$ lies different degrees of probability.

By Bayes's Theorem (p. 42) $P(h/e) = \dfrac{P(e/h)\,P(h)}{P(e)}$ if $P(e) \neq 0$. If e is our present evidence of observation $P(e) \neq 0$. $P(e)$ will be the same for the different hypotheses h about the interval within which $Pr(Q/\alpha)$ lies.

Hence for each such hypothesis h $P(h/e)$ is proportional to $P(e/h)\,P(h)$. Theories about which of a finite number of equal intervals is occupied by $Pr(Q/\alpha)$ are more probable, the more accurate they are, and the greater their intrinsic probability. Since they do not differ from each other in content (they are equally precise theories about the same objects), their relative intrinsic probability will be a matter of relative simplicity. We saw in Chapter VIII that isolated theories of the form '$p_{i-1} < Pr(Q/\alpha) \leqslant p_i$' do not differ from each other greatly in respect of intrinsic probability. Their relative accuracy was the main determinant of the relative probability of such theories. However in view of our wider evidence let us now concern ourselves with theories showing how elementary probabilities such as $Pr(Q/A)$ determine compound probabilities such as $Pr(Q/A \cap B \cap C)$ and so $Pr(Q/\alpha)$. The addition of such factors as 'B', '$B \cap C$', 'C', etc. to 'A' would seem intrinsically as likely to increase the statistical probability of an object which is A being Q as to decrease it (e.g. $Pr(Q/A \cap B \cap C)$ is as likely to be less than $Pr(Q/A)$ as to exceed it). Hence if our sole evidence is that $\Phi(Q/A) = p$ and that a is A, B, C, D etc., $Pr(Q/\alpha)$ would seem as likely to exceed $Pr(Q/A)$ as to fall short of it. For theories which postulate that $Pr(Q/\alpha) > Pr(Q/A)$ will, for a given value assigned to $Pr(Q/\alpha)$, prove as accurate as theories which postulate that $Pr(Q/\alpha) < Pr(Q/A)$, and being intrinsically equally probable are equally probable on the cited evidence. So the value of the probability that a is Q on that evidence will not be very far off p. However we shall often have more evidence than that cited just now. Then we shall need to consider more carefully the relative simplicity of theories showing how elementary statistical probabilities determine compound ones. Such a theory will be simple in so far as it postulates simple interactions between such probabilities, that is simple ways in which elementary probabilities determine compound probabilities. Rival theories on this issue can be constructed. Each theory will consist of claims about the range within which lie $Pr(Q/A)$, $Pr(Q/A \cap B)$, $Pr(Q/A \cap B \cap C)$, $Pr(Q/A \cap C)$ and so on, and theories will differ according to the simplicity of the relationships postulated between these factors. Although criteria of comparative simplicity do not seem to be very clear in this region, certain theories do seem to be markedly simpler than rivals.

These are those theories about the influence of some property C to the effect that it always has a similar influence on the statistical probability of an object being Q, whichever other properties (of those which belong to the conjunction α) the object possesses. In so far as such a theory proves more accurate than any except a much more complicated theory about the influence of C, that makes it much more probable than any rival. One such theory of interaction is the theory that, among all the other properties whose conjunction forms α, one property C is what I shall call totally relevant. By this I mean that for all properties x which belong to α, and for all conjunctions x of such properties, $Pr(Q/C \cap x) \simeq Pr(Q/C)$. Another such theory is the theory that one property, which forms part of the conjunction α, is irrelevant. By the theory that C is irrelevant I mean the theory that for all properties x which belong to α, and for all conjunctions x of such properties, $Pr(Q/x) \simeq Pr(Q/C \cap x)$. Another such simple theory is the theory that C never decreases the statistical probability of an object being Q. This is to say that for all properties x which belong to α, and for all conjunctions x of such properties, $P(Q/C \cap x) \geqslant P(Q/x)$. Such a theory I will call a theory of the positive relevance of C. Another such simple theory is the theory that C never increases the statistical probability of an object being Q. This is to say that for all properties x which belong to α, and for all conjunctions x of such properties, $P(Q/C \cap x) \leqslant P(Q/x)$. Such a theory I will call a theory of the negative relevance of C.

Let us now give examples of situations where such theories are sufficiently more probable than rivals to provide a rough value of the probability that a is Q. We will suppose that we have evidence from very large samples of the values of $\Phi(Q/A)$, $\Phi(Q/B)$, $\Phi(Q/C)$, $\Phi(Q/A \cap B)$, $\Phi(Q/A \cap C)$, and $\Phi(Q/B \cap C)$. 'α' is '$(A \cap B \cap C)$'; that is, all we know about an object a is that it is A and B and C. We wish to know the probability of its being Q. But we have no value of $\Phi(Q/A \cap B \cap C)$, only the evidence so far cited. By the considerations adduced in Chapter VIII it is much more probable that $Pr(Q/A) \simeq \Phi(Q/A)$ than that it has any other value. This seems to remain the case when our evidence is not merely that many A's have been observed and that $\Phi(Q/A)$ has a certain value, but also when the evidence is as cited here. To fill out the example, let us suppose we are inexperienced botanists experimenting with seeds of a new variety of flower. We wish to ascertain in what kind of ground the seeds will take best. Let 'Q' be 'grows into a healthy plant', 'A' = 'is sown in well watered soil', 'B' = 'is sown in sunny soil', 'C' = 'is sown in sandy soil'. We wish to ascertain the probability of a particular seed a producing a healthy plant, when all we know of the seed is that it is planted in well watered, sunny, sandy soil. This probability will be approximately p if it is much more probable that $Pr(Q/A \cap B \cap C)$ is close to p than that it has any other value.

Now suppose, more precisely, that our evidence is that $\Phi(Q/C) \simeq$

$\Phi(Q/A \cap C) \simeq \Phi(Q/B \cap C) = 0.9$, and that $\Phi(Q/A) \simeq \Phi(Q/B) \simeq \Phi(Q/A \cap$ $= 0.1$. Now there are three possible theories about relevant statistical proba bilities, which are exclusive and exhaustive. One is T, that $Pr(Q/A \cap B \cap C$ $\simeq Pr(Q/C) \simeq Pr(Q/A \cap C) \simeq Pr(Q/B \cap C) \simeq 0.9$ (while $Pr(Q/A) \simeq Pr(Q/B)$ $\simeq Pr(Q/A \cap B) \simeq 0.1$). This is an hypothesis of the total relevance of C. Another theory is T', that while $Pr(Q/A) \simeq Pr(Q/B) \simeq Pr(Q/A \cap B) \simeq 0.1$ and $Pr(Q/C) \simeq Pr(Q/A \cap C) \simeq Pr(Q/A \cap B) \simeq 0.9$, $Pr(Q/A \cap B \cap C)$ is not close to 0.9. A third theory is T'', that one or more of $Pr(Q/C)$, $Pr(Q/A \cap C)$, or $Pr(Q/A \cap B)$ do not equal approximately 0.9, or one or more of $Pr(Q/A)$, $Pr(Q/B)$, $Pr(Q/A \cap B)$ do not equally approximately 0.1. Now T'' is highly inaccurate, since we have supposed $\Phi(Q/C)$ etc. to be derived from very large samples; and so, since, intuitively, its intrinsic probability is not vastly in excess of those of the other theories (so long as the 'approximately' is not taken too narrowly) its posterior probability is very low. T', on the other hand, although equally accurate with T, is much less simple than T and so intrinsically much less probable. For in- tuitively it is much more likely, if A and B taken separately have no influ- ence on the statistical probability of a C being Q, that they have no influ- ence taken together, than that they have a considerable influence taken together – especially in view of the fact that in general being a B as well as an A makes an object no more likely to be Q (i.e. $Pr(Q/A \cap B) \simeq$ $Pr(Q/A)$). In this case the hypothesis of total relevance is, I suggest, con- siderably more probable than its rivals, on the cited evidence. So it is considerably more likely that $Pr(Q/A \cap B \cap C) \simeq 0.9$ than that it has any other value, and so the probability on the cited evidence that a is Q is in the region of 0.9. The result is, I suggest, intuitively plausible. If we had to bet on the particular seed of the new variety producing a healthy plant 0.9 to 0.1, or 9 to 1, would seem fair odds.

Suppose however that our more precise evidence is that $\Phi(Q/C) \simeq$ 0.5, that $\Phi(Q/A \cap C) \simeq \Phi(Q/A) = 0.6$, that $\Phi(Q/B \cap C) \simeq \Phi(Q/B) = 0.6$, and that $\Phi(Q/A \cap B) = 0.8$. Here again we can construct three possible theories, of which one is much more probable on the cited evidence than the others. One is T, that $Pr(Q/C) \simeq 0.5$, $Pr(Q/A \cap C) \simeq Pr(Q/A) \simeq 0.6$, $Pr(Q/B \cap C) \simeq Pr(Q/B) \simeq 0.6$, $Pr(Q/A \cap B) \simeq Pr(Q/A \cap B \cap C) \simeq 0.8$. This is an hypothesis of the irrelevance of C. Another is T', that the statis- tical probabilities are the same as in T except that $Pr(Q/A \cap B \cap C)$ is not close to 0.8. The third theory is T'', that at least one of the statistical probabilities other than $Pr(Q/A \cap B \cap C)$ has a value other than that stated in T. Again T'' will be highly inaccurate and so highly improbable. But T', while equally accurate with T, is much less simple and so intrinsically much less probable. Intuitively, if C has no effect on the statistical probability of an A being Q or on the statistical probability of a B being Q, then it is considerably more likely that it has no effect on the probability of an object which is A and B being Q than that it does have such an effect. So it is considerably more likely that $Pr(Q/A \cap B \cap C) \simeq 0.8$ than that it has any other value, and so the probability on the cited evidence that a is Q is in the region of 0.8. The result is again, I suggest, intuitively plausible. The probability of the particular seed producing a healthy plant is in the region of 0.8.

A theory of total relevance and a theory of irrelevance (as also a theory of positive relevance and a theory of negative relevance) will be rendered more probable in so far as they are found to give accurate results for a much larger set of properties than the 'A', 'B', and 'C' considered, 'α' being a conjunction of a much larger set of properties. Thus in so far as C proved totally relevant for many individual properties, pairs of properties and triads of properties x, which belonged to α (i.e. $\Phi(Q/C.x)$ $\simeq \Phi(Q/C)$ for these), the theory that $Pr(Q/\alpha) \simeq \Phi(Q/C)$ is rendered more probable and so the probability that a is Q is given more precisely by $\Phi(Q/C)$. (Among philosophers of science Pap has drawn attention to the use of theories of irrelevance. See [3].)

Let us illustrate the influence of a theory of the positive relevance of C by the following example. Let our more precise evidence be that $\Phi(Q/A) = 0.7$, $\Phi(Q/B) = 0.8$, $\Phi(Q/C) = 0.8$, $\Phi(Q/A \cap B) = 0.9$, $\Phi(Q/A \cap C)$ $= 0.95$, $\Phi(Q/B \cap C) = 0.85$. Let T be the theory that $Pr(Q/A) \simeq 0.7$, $Pr(Q/B) \simeq 0.8$, $Pr(Q/C) \simeq 0.8$, $Pr(Q/A \cap B) \simeq 0.9$, $Pr(Q/A \cap C) \simeq 0.95$, $Pr(Q/B \cap C) \simeq 0.85$, and $Pr(Q/A \cap B \cap C) \geqslant 0.95$. This is an hypothesis of the positive relevance of A and of B and of C. The addition of A to one or other or both of the other properties raises the statistical probability of an object possessing the latter being Q; and the same applies to the addition of B or C. Let T' be the theory that the statistical probabilities are the same as in T, except that $Pr(Q/A \cap B \cap C) < 0.95$. Let T'' be the theory that one or more of the other statistical probabilities given in T are other than as stated there. T'' is highly improbable because highly inaccurate. T' is equally accurate with T but much less probable than it because much less simple. T says that what happens normally – an object's being an A as well as a B or an A as well as a C or a B as well as a C increasing the statistical probability of its being Q – also happens when an object has the third property as well. T' on the other hand states that although $Pr(Q/A \cap B) > Pr(Q/A)$, nevertheless $Pr(Q/A \cap B \cap C) < Pr(Q/A \cap B)$ and so that $Pr(Q/A \cap B \cap \sim C) > Pr(Q/A \cap B)$. This happens, according to T', despite the fact that normally being C increases the statistical probability of an object which is A or B being Q. T' also claims that although $Pr(Q/A \cap C) > Pr(Q/A)$, nevertheless $Pr(Q/A \cap B \cap C) < Pr(Q/A \cap C)$ and so that $Pr(Q/A \cap \sim B \cap C) > Pr(Q/A \cap C)$. This happens despite the fact that being B normally increases the statistical probability of an object which is A or C being Q. And so on. T' is an odd hypothesis, intrinsically much less probable than T. So, I suggest, it is highly probable that $Pr(Q/A \cap B \cap C)$ is at least 0·95, and hence the probability that a is Q on the evidence so far cited is in the region between 1 and 0·95. This result again I suggest we find intuitively to be correct. If we had to bet on the probability of a being Q when all the evidence which we had was that cited above, we would judge it to be very likely that a would be Q.

We would expect many of the properties studied by insurance companies and found by them to increase the statistical probability of death, to be properties which increase the probability of death, whichever other of those properties a man had. We would expect the fact that a man had suffered from cancer, had suffered from tuberculosis, was a frequent air traveller, and a racing driver by profession, properties which each in general increase the statistical probability of death, to be properties which increase the probability of death, whichever other of those properties a man had. We would expect to find in any sample that the frequency of death before a certain age among professional racing drivers who have suffered from cancer to be at least as great as the frequency of death before that age among professional racing drivers as a whole and so at least as great as the frequency of death among professional racing drivers who have not suffered from cancer. But we would also expect, even if such data had not been assembled, that insurance companies would take it for granted, as far more probable on the cited evidence than any alternative, that the statistical probability of death before a certain age by a professional racing driver who had suffered from cancer was at least as great as the probability of death before that age by a professional racing driver or a professional racing driver who had not suffered from cancer.

If the less simple theory in the above cases were more accurate – e.g. if statistical data $\Phi(Q/A \cap B \cap C)$ were predicted more accurately by it than by the simpler theory – that would make the more complicated supposition more likely to be true; but in the absence of statistical data the simpler supposition is, I suggest, much more likely to be true. Such theories as a theory of negative or positive relevance are very imprecise theories about how elementary (complex statistical) probabilities interact to determine compound ones. They do not tell us how the elementary

probabilities determine an approximate value of a compound one, only
that the compound one has a value greater or less than some precise value.
To the extent that a more detailed theory of interaction is simpler and
proves more accurate than the disjunction of rival theories, it is thereby
shown to be more probable than all those rivals taken together, and so
can be used to give a more precise value to the probability that a is Q.
Thus it could be postulated of a certain class of properties that for any
three such properties x_i, x_j, and x_k, always

$$Pr(Q/x_i \cap x_j \cap x_k) \simeq \frac{Pr(Q/x_i \cap x_j)\, Pr(Q/x_i \cap x_k)}{Pr(Q/x_i)}.$$

The claim that for particular x_i, x_j, and x_k, $Pr(Q/x_i \cap x_j \cap x_k)$ had the
value stated by this theory rather than that it had any other value would
be rendered more probable in so far as the theory yielded accurate predictions of statistical data for many members of the class of properties,
predictions more accurate than those of any theory of comparable simplicity. However for it to be *much* more probable in a particular case that
$Pr(Q/x_i \cap x_j \cap x_k)$ had the value stated than that it had any other value,
and so for us to be able to say that the probability of an object a, known
to be x_i, x_j, x_k, being Q had that value, we should need a lot of statistical
data.

If a quantifiable dependent variable – e.g. age – is involved we can
construct theories showing the probability of some characteristic – e.g.
death – for each value of the variable. The theories will have the form

$$Pr(Q/I_x) = f(x) \pm r.$$

Where I_x means that the value of the variable I is x, $f(x)$ is a certain
function of x, and r is a constant. Such theories are more probable to
the extent to which they are simple and accurate. We can, as we saw in
Chapter VII, to some extent rank formulae $f(x)$ in order of simplicity.
A linear function of x is simpler than a quadratic or a logarithmic function, a quadratic than a cubic, etc. To find formulae of this type is a
regular task for the applied statistician. Textbooks of applied statistics
recommend trying formulae of linear before curvilinear equations, and
they then list simple formulae of curvilinear equations which it would be
useful for the investigator to try to fit to the data (see e.g. [9] p. 70).
The discovery of such formulae is crucial for projection of time trends.

The rules which I have stated so far in this chapter and in Chapter
VIII for calculating how elementary statistical probabilities such as
$Pr(Q/A)$ interact to determine the value of a compound statistical probability $Pr(Q/\alpha)$ only operate under normal circumstances. By this I mean

that they only operate when there are no peculiar logical relationships between 'Q', 'A', and 'α' which show strong reason for postulating a special kind of interaction between the statistical probabilities involved. This will happen in two kinds of case. The first is where $Pr(Q/\alpha)$ or $Pr(Q/A)$ can be assigned a value on logical grounds alone. Thus if 'Q' is 'α', then $Pr(Q/\alpha) = 1$, whatever the value of $Pr(Q/A)$. Or if 'A' is 'member of a species, the majority of the members of which are Q', necessarily $Pr(Q/A) > \frac{1}{2}$.

More interesting is the kind of case where, although $Pr(Q/\alpha)$ or $Pr(Q/A)$ cannot be assigned a value on logical grounds alone our criteria of epistemic probability show good grounds in virtue of the logical relations between 'Q', 'A', and 'α' for supposing that the relation between the cited statistical probabilities is different from the normal one. This will be the case where one or both of '$Pr(Q/\alpha)$' and '$Pr(Q/A)$' can be rewritten in terms of logically equivalent simpler predicates, and where rules such as those given earlier make probable a certain relationship between the rewritten statistical probabilities which was not revealed by the original form in which they were expressed. I understand by 'B', 'C', etc. being simpler predicates than 'A', 'Q' etc., that universal nomological propositions which use the former are by the criteria of Chapter VII in general simpler than ones using the latter predicates, e.g. '$\Pi(B/C) = 1$' is simpler than '$\Pi(Q/A) = 1$'. When probabilities can be expressed in terms of simpler predicates this should be done before rules such as those given earlier in this and in the previous chapter are applied. They are meant to apply only when the statistical probabilities have been expressed in terms of the simplest predicates possible. I will give an example of the application of this consideration very shortly when we come to consider a problem of a slightly different type from those considered so far in this chapter.

So far in this chapter we have considered how the probability that a is Q is to be calculated on evidence that it is α and evidence about the proportion of Q's in wider classes than α. We must now consider, though we must do so very briefly, how the probability that a is Q is to be calculated on evidence that it is α and evidence about the proportion of Q's in narrower classes than α. Exactly the same considerations apply here as with problems of the earlier type. We use the evidence e to assess the probabilities of different theories about the value of $Pr(Q/\alpha)$ and then use the weighted sum formula to calculate the value of the probability on e that a is Q.

For example, suppose we know that a is α, that many α's all of which are G have been observed and that $\Phi(Q/\alpha \cap G) = q$. Then it is highly probable that $Pr(Q/\alpha \cap G) \simeq q$. We saw earlier that theories attributing to $Pr(Q/\alpha \cap G)$ a value higher than $Pr(Q/\alpha)$ were intrinsically equally probable with theories attributing to $Pr(Q/\alpha \cap G)$ a value lower than $Pr(Q/\alpha)$. Hence we may conclude that the probability that a is Q is roughly q. Suppose that if we have only examined α's (e.g. university teachers aged under fifty) who are also G (e.g. live in Hull) to investigate the probability of their being Q (e.g. dying before the age of sixty) and found that $\Phi(Q/\alpha \cap G) = 0.05$. We know of another individual a only that he is α (a university teacher aged under fifty) but not whether he is G or $\sim G$. Then in the absence of further information (e.g. about the influence of G or similar properties on other statistical probabilities) we can conclude that the probability that a is G is roughly 0.05. (Of course often we will have background evidence about the probable effect of G, but I am assuming in this case that we do not.)

However this conclusion, like all earlier conclusions of these last two chapters, is subject to the qualification that the predicates concerned cannot be expressed in terms of simpler predicates by the criteria stated on p. 148. On evidence that many emeralds have been observed before AD 2000 and all found to be grue, we cannot conclude that the probability of an emerald a at an unknown date being grue is close to 1. (For the definition of 'grue' see p. 101.) This is because the evidence can be expressed in terms of the predicate 'green' which by the stated criteria is markedly simpler than 'grue'. From that evidence we can conclude that the probability of a being green is close to 1. Hence if we suppose that intrinsically an emerald is as likely to exist after AD 2000 as before AD 2000, we can conclude that the probability of a being grue is somewhere in the region of 0.5.

So then, I would claim, there are rational procedures, which I have set forward in outline, for calculating, at any rate roughly, a range within which $Pr(Q/\alpha)$ is much more likely to lie than to lie anywhere else, and so a rough value for the probability that a is Q – given that the only available data are those of the observed proportions of Q's among objects having properties possessed by a.

We will normally have wider evidence than the statistical data just referred to, which affects the value of the probability of a particular proposition. We will have evidence from a very wide field about what kinds of properties are likely to make a difference to other kinds of properties – e.g. that the letter with which a man's surname begins is unlikely to affect his health; that the fact that the year in which it was born is divisible by three is unlikely to affect whether a horse wins a race, and so on. This very general evidence will make some theories of how elementary (complex statistical) probabilities interact to affect compound ones more probable than others. We have not considered in this chapter the way in which such evidence affects the probability of such theories. Our discussion has assumed a very limited kind of evidence, and has attempted to answer the question how the probability of a particular proposition is to be calculated on such evidence. The influence of wider evidence will be considered briefly in Chapter XI.

The results of the last two chapters were derived from very general considerations about evidence, but the procedures set out here are, I believe, very broadly those followed by insurance companies in calculating different premium rates for different individuals. (The procedures used by actuaries in calculating mortality tables for the use of insurance companies are set out in a work such as [8] – see especially Chapter XIII of that work summarized on pp. 146f. of the work.)

Bibliography

For standard accounts by philosophers of how the probability of particular propositions is to be calculated on evidence of the kind considered in this chapter, see:

[1] H. REICHENBACH *The Theory of Probability*, Berkeley and Los Angeles, 1949, Chapters 9 and 11.

[2] RUDOLF CARNAP *The Continuum of Inductive Methods*, Chicago, 1952.

[3] A. PAP *An Introduction to the Philosophy of Science*, New York, 1962, pp. 186-9.

[4] A. J. AYER 'On the Probability of Particular Events' in *The Concept of a Person*, London, 1963.

[5] WESLEY C. SALMON *The Foundations of Scientific Inference*, Pittsburgh, 1967, pp. 83-108.

For criticism of Carnap's account, see:

[6] P. ACHINSTEIN 'Variety and Analogy in Confirmation Theory' *Philosophy of Science*, 1963, **30**, 207-21.

[7] M. HESSE 'Analogy and Confirmation Theory' *ibid.*, 1964, 31, 319–27.

For works of applied statistics cited in the chapter, see:

[8] J. L. ANDERSON and J. B. DOW *Actuarial Statistics. Volume II. Construction of Mortality and other Tables,* Cambridge, 1948.

[9] M. EZEKIEL and K. A. FOX *Methods of Correlation and Regression Analysis,* 3rd edition, New York, 1959.

CHAPTER X

Nicod's Criterion

The results of Chapter VIII will be useful in helping us to solve a further problem about comparative probability – the worth of Nicod's criterion.

Nicod's criterion, we saw in Chapter IV, states that (for any background evidence k) '$\phi a . \psi a$' always confirms, '$\phi a . \sim\psi a$' always disconfirms, while '$\sim\phi a . \psi a$' and '$\sim\phi a . \sim\psi a$' are always irrelevant to, 'of physical necessity all ϕ's are ψ'. We argued in Chapter IV that the paradoxes of confirmation showed that Nicod's criterion was false. That the cause of the paradoxes is the mistaken adoption of Nicod's criterion has indeed been the view of the majority who have written on this topic. Yet since Nicod's criterion is intuitively plausible, those who adopt the solution of denying it need to explain why we are wrongly inclined to adopt it and to make clear to our intuition its implausibility. Different writers have however given different accounts of this matter and have also provided different alternatives to Nicod's criterion. This chapter will consider further arguments (other than the argument that it generates the paradoxes of confirmation) for the falsity of Nicod's criterion, and will consider the true confirmation relations of '$\phi a . \psi a$', '$\phi a . \sim\psi a$', '$\sim\phi a . \psi a$', and '$\sim\phi a . \sim\psi a$' to 'of physical necessity all ϕs are ψ'.

Let us denote '$\phi a . \psi a$' by 'b_1', '$\phi a . \sim\psi a$' by 'b_2', '$\sim\phi a . \psi a$' by 'b_3' and '$\sim\phi a . \sim\psi a$' by 'b_4', and 'of physical necessity all ϕ's are ψ' by 'h'. Now one part of Nicod's criterion is undoubtedly correct – given two assumptions. Given that $P(b_2/k) \neq 0$ and that $P(h/k) \neq 0$, b_2 always disconfirms h. For by Theorem 1 of the calculus (p. 40) $P(\sim b_2/h . k) = 1$ and so by Theorem 2 (p. 40) $P(b_2/h . k) = 0$. Hence $P(b_2/h . k) < P(b_2/k)$. Hence by Principle C (p. 48) b_2 disconfirms h. By the argument of p. 43 $P(b_2/k)$ will not equal 0 unless b_2 is incompatible with k, since b_2 is the report of a particular observation. If 'ϕ' and 'ψ' are both projectible predicates (e.g. are not positional predicates – see Chapter V for discussion of which predicates are projectible), then $P(h/k) \neq 0$. I shall henceforward assume that k is compatible with b_2 and that h uses projectible predicates. On these assumptions it is undoubtedly correct that b_2 always disconfirms h. I shall also assume that k is compatible with b_1, with b_3, and, with b_4, as well as with b_2, while not entailing any of them. I assume, that is, that the evidence available so far is compatible with but does not entail any of the possible observations. Hence for each b, $0 \leqslant P(b/k) \leqslant 1$.

It might also seem to follow with these assumptions from Principle C that '$\phi a . \psi a$', that is 'b_1', always confirms h, 'of physical necessity all ϕ's are ψ'. But it does not. $P(\psi a/h . k . \phi a) = 1$. If k includes ϕa $P(b_1/h . k) = 1$ and so given that $P(b_1/k) \neq 1$, $P(b_1/h . k) > P(b_1/k)$ and so by Principle C b_1 confirms h. But unless 'ϕa' is supposed to be included in the background evidence k, $P(b_1/h . k)$ does not in general equal 1 and so the argument does not work. The confirmation relation of b_1, as of b_3 and b_4, to h is not so simple a matter.

Among those who deny Nicod's criterion we may distinguish three alternative approaches. I will discuss each in turn. A plausible account of the relation between them is given by Mackie [5], though we shall see that in certain crucial details Mackie's account is mistaken.

I will begin by considering the account given by Hempel [1] who was the original opponent of Nicod's criterion. According to Hempel 'b_1', 'b_3', and 'b_4' all confirm h (which I shall henceforward write simply as 'all ϕ's are ψ'). Hempel's main argument for this conclusion seems to be that, assuming that '$\phi a . \psi a$' always confirms 'all ϕ's are ψ' (something which he takes for granted), the wider conclusion is a consequence of the scientific laws condition and the c-condition. (See Chapter IV for these.) For by the scientific laws condition h 'all ϕ's are ψ' is logically equivalent to h' 'everything which is (ϕ or $\sim\phi$) is (ψ or $\sim\phi$)'. By the c-condition whatever confirms h' confirms h. But, Hempel argues, as '$\phi a . \psi a$' always confirms 'all ϕ's are ψ', so similarly '(ϕ or $\sim\phi$)a. (ψ or $\sim\phi$)a' that is, either '$\phi a . \psi a$' or '$\sim\phi a . \psi a$' or '$\sim\phi a . \sim\psi a$' will always confirm h' and so h.

The conclusion seems however somewhat paradoxical. Does the discovery of every non-raven, black or non-black, really confirm 'all ravens are black'? So Hempel gives two subsidiary arguments designed to show that the paradoxicality is only apparent. The first is this (see [1] p. 18). We are falsely inclined to suppose that 'all ϕ's are ψ' is only about ϕ's. Certainly our interest in the hypothesis may arise solely from our interest in ϕ's, but the hypothesis says something about every object x in the world, that it is not a case of $\phi x . \sim\psi x$. Hence, Hempel would say (though he does not do so explicitly in this passage) one would expect evidence about any object x that it is not a case of ϕx. $\sim\psi x$ to confirm 'all ϕ's are ψ'.

His second subsidiary argument is this (see [1] p. 19). Our inclination to suppose that a piece of ice held in a flame and found not to burn yellow does not confirm 'all sodium salt burns yellow' arises from our supposing that we already know that it is a piece of ice. In that case the hypothesis places no further restrictions on its behaviour – its burning yellow or not burning yellow are both in accord with the hypothesis. Hence one result rather than the other is not going to give support to the hypothesis. But suppose that we first find that an object does not burn yellow and then we discover that it is ice. The latter discovery does confirm the hypothesis, because a different discovery – that it was sodium salt – could have overthrown the hypothesis. The apparent paradoxicality of '$\sim\phi a$. $\sim\psi a$' confirming 'all ϕ's are ψ' arises from our supposing that '$\sim\phi a$. $\sim\psi a$' reports our discovery about a previously identified $\sim\phi$ that it is $\sim\psi$. But it does not. It reports our discovery about an object, not previously identified by the possession or lack of ϕ-ness or ψ-ness, that it is $\sim\phi$ and $\sim\psi$. When we realize this we see that '$\sim\phi a$. $\sim\psi a$.' does confirm the hypothesis, since it shows that a certain object is not a falsifying instance of it.

It will be seen that central to both of these subsidiary arguments of Hempel to explain away the apparent paradoxicality of '$\sim\phi a$. ψa' and '$\sim\phi a$. $\sim\psi a$' confirming 'all ϕ's are ψ' are two assumptions, without which the arguments would have no force. These are:

(A) the discovery that an object is not a falsifying instance of an hypothesis confirms the hypothesis;

(B) if an observation is the result of a test, no result of which could, with or without other evidence, falsify a certain hypothesis, then that observation cannot confirm the hypothesis.

(A) is clearly taken for granted in both arguments. (B) is implicit in the second argument, for instance, when Hempel writes that once we know '(i) that the substance used in the experiment is ice, and (ii) that ice contains no sodium salt . . . then of course the outcome of the experiment (viz. finding which colour the substance burns, if it burns) . . . can add no strength to the hypothesis under consideration' (viz. the hypothesis 'all sodium salts burn yellow') ([1] p. 19). Without these two assumptions the subsidiary arguments carry no weight.

Assumption (A) seems to me plausible but irrelevant. The discovery that a certain object a is not a falsifying instance of 'all ϕ's are ψ' is the discovery that '$\sim\phi a$ v ψa'. It is plausible to suppose that this counts in favour of the hypothesis. But what is at stake is the effect of more specific information, e.g. '$\sim\phi a . \sim\psi a$', on the hypothesis. The more specific information might count the other way. For instance, the discovery that some object other than a ϕ was $\sim\psi$ might suggest that ϕ's too were likely to be $\sim\psi$. As containing '$\sim\phi a$ v ψa', '$\sim\phi a . \sim\psi a$' counts for the hypothesis, but other information contained within it might count against the hypothesis, so on balance '$\sim\phi a . \sim\psi a$' might be irrelevant or disconfirmatory. Assumption (B) is clearly false as a general principle. Observations which result from tests, no result of which could falsify an hypothesis, may sometimes confirm the hypothesis. An obvious case arises when the hypothesis is statistical – e.g. that $Pr(B/A)$ lies between 1 and 0·95. Finding an A to be B confirms the hypothesis, although no observation could falsify the hypothesis.

I conclude that neither of Hempel's two subsidiary arguments succeeds in showing that the paradoxicality of his original conclusion, that '$\phi a . \psi a$', '$\sim\phi a . \psi a$', and '$\sim\phi a . \sim\psi a$' all confirm 'all ϕ's are ψ', is only apparent. Further, one of the points which I made in rebutting these arguments shows Hempel's main argument to be fallacious. Hempel assumed that '$\phi a . \psi a$' always confirms 'all ϕ's are ψ'. He therefore argued that '(ϕ or $\sim\phi$)a. (ψ or $\sim\phi$)a' confirms 'everything which is (ϕ or $\sim\phi$) is (ψ or $\sim\phi$)'. Hence, by the c-condition, '(ϕ or $\sim\phi$)a. (ψ or $\sim\phi$)a', that is '$\sim\phi a$ v ψa' confirms 'all ϕ's are ψ'. However Hempel wrongly concluded that if '$\sim\phi a$ v ψa' confirms 'all ϕ's are ψ' then so does any more detailed evidence compatible with '$\sim\phi a$ v ψa', e.g. '$\sim\phi a . \sim\psi a$'. That does not follow. Even if '$\sim\phi a$ v ψa' counts for the hypothesis, nevertheless, as we have seen, the further evidence about a might mean, for anything that Hempel has shown, that on balance '$\sim\phi a . \sim\psi a$' counts against 'all ϕ's are ψ'. So Hempel's argument does not prove his conclusion, even if we grant his assumption that '$\phi a . \psi a$' always confirms 'all ϕ's are ψ' (an assumption which we shall call into question later in this chapter).

Hempel does not mention in his discussion the possibility of any background evidence, presumably because he considers that 'b_1', 'b_3', and 'b_4' confirm, and 'b_2' disconfirms 'all ϕ's are ψ' in any circumstances whatsoever. Some writers, however, while accepting that 'b_1' and 'b_4' and possibly 'b_3' always confirm, have urged that we can both show this and understand why Nicod's criterion *seems* correct if we take into account the normal minimum background evidence which we must have when considering such hypotheses as 'all ravens are black'. They suppose that we have at any rate some prior view about the proportion of ravens and of black objects in the world. We could not after all understand the

L

hypothesis without having made some observations of the world. And if we do understand the hypothesis we must know of at any rate some of the objects which we observed whether or not they were black or ravens. So we must have acquired some evidence about the proportion of ravens and of black objects in the world.

Once we begin to talk about proportions of ravens and of black objects in the world, we do of course presuppose the existence of some principle for counting ravens, non-ravens, black objects, and non-black objects. It is clear how to count the number of ravens in some region, e.g. my room. But how many non-ravens are there in the room, how many black objects, how many non-black objects? There are various principles for counting these objects, which could be adopted. We could for example count every piece of matter of raven-size as one object. Or we could count every object of a type for which we had a name as one object (adding rules for preventing us counting both parts and whole of an object). It does not matter much what principle is used. On any reason able principle of counting objects, the crucial mathematical assumptions to be discussed will hold.

I will call the argument to be considered to show the falsity but apparent plausibility of Nicod's criterion the quantitative argument. It was originally put forward by Hosiasson–Lindenbaum [2] and subsequently by Pears [3] and Alexander [4]. A simple and rigorous presentation of Alexander's argument is given by Mackie [5]. In the view of these writers both '$\phi a . \psi a$' and '$\sim\phi a . \sim\psi a$' confirm 'all ϕ's are ψ'. (We will come to the case of '$\sim\phi a . \psi a$' later.) However one of these is liable to confirm the hypothesis much more than the other. How much each will confirm depends on our background evidence about the proportion of ϕ's and of ψ's in the world. Let our background evidence be that the ratio of ϕ's to $\sim\phi$'s is $x:1-x$, and of ψ's to $\sim\psi$'s, $y:1-y$, then, an argument to be given will show, in so far as $x < 1-y$, '$\phi a . \psi a$' will confirm 'all ϕ's are ψ' better than does '$\sim\phi a . \sim\psi a$'. In the case of most hypotheses $x \ll 1-y$ (that is, x is much less than $(1-y)$). Consider for example the hypothesis 'all ravens are black'. Our background evidence is that there are far more non-black things than ravens in the world. Hence '$Ra . Ba$' will confirm 'all ravens are black' much more than does '$\sim Ra . \sim Ba$'. Since our background evidence will be that $x \ll 1-y$ for most hypotheses put forward of the form 'all ϕ's are ψ', '$\phi a . \psi a$' will generally confirm much better than does '$\sim\phi a . \sim\psi a$'. This will make us falsely inclined to believe that '$\sim\phi a . \sim\psi a$' does not confirm 'all ϕ's are ψ' at all. Hence the source of the plausibility of Nicod's criterion. Yet if the claim made in this paragraph is correct, the criterion is false.

The argument to the claim of the last paragraph about the relative strength of confirming instances begins with our Principles C and D (pp. 48ff.) (parts of Mackie's relevance criterion), which state, in Mackie's summary, that 'a hypothesis h is confirmed by an observation report b in relation to background knowledge k if and only if . . . $P(b/k \cdot h) > P(b/k)$ and that b confirms h the better the more the adding of h to k raises the probability of b' ([5] p. 167). (The main point of Mackie's paper was to show the dependence of various solutions on his relevance criterion rather than to defend Alexander's particular solution, but because it is clear and has been much discussed I choose to develop Mackie's account of the latter.) Now suppose that our hypothesis h is as before 'all ϕ's are ψ' and our background evidence which I shall in this case call K_0, is that the ratio of ϕ's to $\sim\phi$'s is $x:1-x$, and of ψ's to $\sim\psi$'s is $y:1-y$ (and so both $0 \leqslant x \leqslant 1$ and $0 \leqslant y \leqslant 1$). ($K_0$, that is, consists of two propositions of simple statistical probability.) Then, the argument goes, the probabilities of an observation of each kind on the different evidence are as follows:

TABLE I

	$b = b_1(\phi a \cdot \psi a)$	$b = b_2(\phi a \cdot \sim\psi a)$	$b = b_3(\sim\phi a \cdot \psi a)$	$b = b_4(\sim\phi a \cdot \sim\psi a)$
$P(b/K_0)$	xy	$x(1-y)$	$y(1-x)$	$(1-x)(1-y)$
$P(b/K_0 \cdot h)$	x	0	$y-x$	$1-y$

The probabilities given in the first line seem intuitively obvious, and the advocates of the quantitative approach do not give any argument for them. But it should be noted that these probabilities will not hold universally whatever 'ϕ' and 'ψ' represent. By Axiom IV of the calculus $P(b_1/K_0) = P(\phi a \cdot \psi a/K_0) = P(\psi a/K_0 \cdot \phi a) \times P(\phi a/K_0)$. Now $P(\phi a/K_0)$ must equal x. If all we know is what proportion of objects are ϕ and ψ, then the probability of a particular object, a, being ϕ, is surely given by the proportion of ϕ's in the total population. (See p. 29.) What of $P(\psi a/K_0 \cdot \phi a)$? Advocates of the quantitative approach have assumed that $P(\psi a/K_0 \cdot \phi a) = P(\psi a/K_0) = y$. Since K_0 includes only the quantitative data cited, it cannot include empirical evidence that being a ϕ makes an object more likely to be a ψ than it would otherwise be. So if being a ϕ makes an object any more likely to be a ψ than it would otherwise be, it must be in virtue of the meaning of 'ϕ'. 'ϕ' and 'ψ' may have such meanings that, either of logical necessity or in virtue of our criteria of epistemic probability, $P(\psi a/K_0) \neq P(\psi a/K_0 \cdot \phi a)$. Let '$\phi$' be 'mammal belonging to a species most of whose members are ψ'. In that case even if $P(\psi a/K_0) = y < \frac{1}{2}$, $P(\psi a/K_0 \cdot \phi a) > \frac{1}{2}$, and so $P(\psi a/K_0) \neq P(\psi a/K_0 \cdot \phi a)$.

However if 'ϕ' does not have a peculiar meaning of this kind which makes an object which is ϕ any more or less likely to be ψ than otherwise, then $P(\psi a/K_0) = P(\psi a/K_0 . \phi a)$. For normal predicates such as 'ϕ' = 'raven' and 'ψ' = 'black', $P(\psi a/K_0)$ does seem to equal $P(\psi a/K_0 . \phi a)$. In that case $P(\phi a . \psi a/K_0) = xy$. Similar argument will justify the other figures in the first line.

The probabilities in the second line are reached as follows. If all ϕ's are ψ, then the probability of observing a ϕ which is $\sim\psi$ will be 0, and the probability of observing a ϕ which is ψ will be the same as the probability of observing a ϕ, viz. x. If all ϕ's are ψ, then no ϕ's will be $\sim\psi$, and so the probability of observing a $\sim\psi$ which is $\sim\phi$ will be the same as the probability of observing a $\sim\psi$, viz. $(1 - y)$. Since the probabilities of all alternatives must sum to 1, the probability of observing a $\sim\phi$ which is ψ will be $(y - x)$. It follows that '$\phi a . \psi a$' is $(1/y)$ times more probable given $(h . K_0)$ than given merely h, and that '$\sim\phi a . \sim a$' is $(1/(1 - x))$ times more probable given $(h . K_0)$ than given merely h. If, as was argued in the last paragraph, it will normally be the case that $x \ll 1 - y$ then (by Principle D) the b_1 observation will confirm much better than the b_4 observation. (The only assumption about proportions necessary for this conclusion is ($x \ll 1 - y$. For h to be compatible with K_0, $x \leqslant y$; if all ϕ's are ψ's, there must be at least as many ψ's as ϕ's. Mackie made the assumption $x < y <$ There is no need for this assumption which I have replaced by the other one.)

This solution unfortunately has the awkward consequence that

'$\sim\phi a . \psi a$' *lowers* the probability of the hypothesis by $\dfrac{y - x}{y - xy}$. This point is very much stressed by Hooker and Stove [14] who emphasize the paradoxicality of supposing that discovery of a black shoe disconfirms 'all ravens are black', and claim for this reason that the solution does not work. It should be noted that this paradoxical (and un-Hempelian) conclusion follows, whatever the values of x and y (whether or not $x \ll 1 - y$ unless $y = 1$ (and in this case h would be entailed by the background evidence k, and so no new evidence could confirm h). The same applies given the argument outlined above to the conclusion that '$\sim\phi a . \sim\psi a$' as well as '$\phi a . \psi a$' confirms 'all ϕ's are ψ'. The assumption that $x \ll 1 - y$ only affects the extent to which, not the fact that, '$\sim\phi a . \sim\psi a$' and '$\phi a . \psi a$' confirm, and '$\sim\phi a . \psi a$' disconfirms 'all ϕ's are ψ'.

The paradoxicality of the result seems to arise because the background knowledge in a normal inductive situation is not of the type described above. Following Alexander, Mackie seems to suppose that the background knowledge will be K_0, that the cited ratios hold among all objects, observed and not yet observed. Sometimes our evidence could be of this kind, as in one of the examples which Alexander cites – 'Suppose we are visiting an American campus where we know that 80 per cent of the male students are undergraduates (ϕ) and that 90 per cent of the male students have short hair (ψ). We then consider the hypothesis that all male undergraduates have short hair – that all ϕ objects are ψ' ([4] p. 231). In this example we know exactly what the proportion of ϕ's and ψ's in the population is; the campus authorities have presumably told us. This makes it reasonable to suppose that a ψ being $\sim\phi$ lessens the probability of all ϕ's being ψ. For a ψ being $\sim\phi$ makes it more likely that other ψ's are $\sim\phi$. But the more likely ψ's are to be $\sim\phi$, then, the proportion of ψ's being fixed, the more probable it is that the proportion in the population of ψ's which are ϕ is less than the number x, the known proportion of ϕ's, and so that not all ϕ's are ψ. Every ψ which is $\sim\phi$ leaves one less to be ϕ. But this is not the normal inductive situation of interest to science. In the normal situation, when we are dealing with a nomological proposition, a 'general empirical hypothesis' of the type which concerned Hempel, the background evidence is going to be k_0, that the cited ratios hold among objects so far observed. We do not *know* what is the proportion in the universe of ϕ's to $\sim\phi$'s, ψ's to $\sim\psi$'s, but we do have evidence on which to base different guesses, some of which will be more probable than others. k_0 is evidence rendering probable to different degrees different hypotheses of complex statistical probability $K_0, K_1, K_2 \ldots$ of the different relations between being ϕ or $\sim\phi$, being ψ or $\sim\psi$. I shall show shortly that when the background knowledge is k_0 rather than K_0 we get a different set of figures from those in Table I. I thus develop a point made by Good ([6] and [7]) that the conclusions that a b_1-observation and a b_4-observation confirm and a b_3-observation disconfirms h follow given K_0 but do not in general follow given a probabilistic distribution over K_0, K_1, K_2, K_3, etc., which is what one will have, given k_0.

It was shown by Hooker and Stove [14] that given Principle C and

(a) $P(Ra/k) = x$

(b) $P(Ba/k) = y$

(c) $P(Ba/k \cdot Ra) = P(Ba/k)$

(d) $P(Ra/k \cdot h) = P(Ra/k)$

(e) $P(Ba/k \cdot h) = P(Ba/k)$

(f) $P(Ba/k \cdot h \cdot Ra) = 1$

we get the result that '$\sim Ra \cdot Ba$' disconfirms. Assuming (f), Principle C, and that '$Ra \cdot Ba$' and '$\sim Ra \cdot \sim Ba$' confirm, they show that one can only avoid the paradoxical result by denying both (d) and (e) and claiming that (relative to k) 'Ra' disconfirms and 'Ba' confirms 'all ravens are black'. This result they describe as a 'violent double new paradox' ([14] p. 313). However, the result can, I believe, be shown to be plausible, and I shall attempt to show its plausibility in the process of showing the consequences of the more realistic assumption about background evidence.

Our background evidence then is k_0, that among objects observed the ratio of ϕ's to $\sim\phi$'s is $x:1 - x$, and of ψ's to $\sim\psi$'s $y:1 - y$. From this we have to infer the probability of a future ϕ being ψ etc. We have examined the problem of this kind of inference in Chapter VIII. We saw that where the sole evidence is that the observed proportion of A's which are Q is q ($\Phi(Q/\alpha) = q$), the probability p on this evidence of a, a future α, being Q is, to a degree of approximation which we can normally ignore, q. However this will only hold if the sample studied is large, and if q is not 1 or 0. In the cases with which we are now concerned the sample is large for it consists of all the objects which we have ever observed. If q is 1 or 0, the probability of a being Q will be close to, but not exactly, 1 or 0 respectively. Bearing in mind this qualification, we can conclude that the probability of a certain future object being ϕ on evidence that the proportion of ϕ's in the sample is x, is x. The additional evidence that the proportion of ψ's in the sample is y would not seem to affect this probability – at any rate, given that being ϕ does not in virtue of the meaning of 'ϕ' make an object more or less likely to be ψ than it would otherwise be. He at any rate for predicates which are not thus interrelated, $P(\phi a/k_0) = x$, $P(\sim\phi a/k_0) = (1 - x)$, $P(\psi a/k_0) = y$, and $P(\sim\psi a/k_0) = (1 - y)$. What can we conclude from this about the probability of compound observations, an object being both ϕ and ψ etc? Exactly the same considerations seem to apply as in the case considered earlier where the background evidence was K_0. $P(b_1/k_0) = P(\phi a . \psi a/k_0) = P(\phi a/k_0) \times P(\psi a/k_0 . \phi a)$. $P(\phi a/k_0)$ we have seen is x. What of $P(\psi a/k_0 . \phi a)$? This, as we argued in the similar case considered before, surely equals $P(\psi a/k_0)$ – unless 'ϕ' has a peculiar meaning which makes an object which is ϕ more or less likely to be ψ than it would otherwise be. Hence, for predicates not thus interrelated, $P(b_1/k_0) = xy$, and by similar arguments $P(b_2/k_0) = x(1 - y)$, $P(b_3/k_0) = (1 - x)y$, and $P(b_4/k_0) = (1 - x)(1 - y)$. We get the same results for the observations on evidence k_0 as on evidence K_0.

But now h, 'all ϕ's are ψ', is added to our evidence. $h . k_0$ tells us two certain things. The first concerns the distribution of ϕ-ness and ψ-ness among the objects which we have so far observed. It is that a proportion x of them were ($\phi \cap \psi$), none were ($\phi \cap \sim\psi$), a proportion ($y - x$) of them were ($\sim\phi \cap \psi$), and a proportion ($1 - y$) of them were ($\sim\phi \cap \sim\psi$). The second is that among objects as yet unobserved none are ($\phi \cap \sim\psi$).

What can we conclude from these two pieces of information about the probabilities of objects being ϕ or ψ in future? The first piece of information allows us to infer, as argued above, that the probabilities of unobserved objects having the various characteristics are (for background evidence k_0 instead of K_0) as indicated in the second line of Table I, and so that there are likely to be less b_3 instances than, without h, we would have supposed. The second piece of information allows an inference which somewhat counters the effect of the first. Evidence that all ϕ's as yet unobserved are ψ is evidence that in this sample consisting of ϕ's from the population of unobserved objects, all members of the sample are ψ. This, by the principles set out in Chapter VIII, is in the absence of other background evidence confirming evidence for hypotheses $Pr(\psi/w)$ $= p$ (w being any unobserved object at all) with high values of p, and so raises the probability of an object as yet unobserved, a $\sim\phi$ as well as a ϕ, being ψ. This probability will be higher in so far as the background evidence contains information that ϕ's are similar to $\sim\phi$'s in their possession of properties like ψ or $\sim\psi$. If the background evidence is that ϕ's are often dissimilar to $\sim\phi$'s in this respect, the fact that all ϕ's are ψ could make it less likely that a $\sim\phi$ was ψ. If all future ϕ's are ψ, then all future $\sim\psi$'s will be $\sim\phi$. This is evidence that in the sample consisting of all $\sim\psi$'s, as yet unobserved, all objects are $\sim\phi$, which by the above argument in the absence of further background evidence raises the probability of an object as yet unobserved, a ψ as well as a $\sim\psi$, being $\sim\phi$. This probability will be higher in so far as the background evidence contains information that $\sim\psi$'s are often similar to ψ's in their possession of properties like ϕ or $\sim\phi$ (and lower if there is evidence that $\sim\psi$'s are often dissimilar to ψ's in their possession of properties like ϕ or $\sim\phi$).

Let us for the moment assume as we have done that the background evidence k_0 consists only of information about the proportions of ϕ's and ψ's not about their similarities in the kinds of way referred to above. Let us see in this case how the two pieces of information work together to determine the values of $P(b/h \cdot k_0)$. Let x_1 be the probability, given $h \cdot k_0$ of an object being ϕ, which, in view of h, will be the same as the probability of a b_1-observation. Let y_1 be the probability, given $h \cdot k_0$, of an object being ψ. Then $(1 - y_1)$ will be the probability, given $h \cdot k_0$, of an object being $\sim\psi$, which, in view of h, will be the same as the probability of a b_4-observation. Given $h \cdot k_0$, the probability of a b_2-observation will be 0, and hence, given $h \cdot k_0$, the probability of a b_3-observation, since probabilities sum to 1, must be $y_1 - x_1$. Hence the table showing the probability of the different observations on different evidence k_0 and

$k_0 . h$ should read as follows:

TABLE II

	$b_1(\phi a . \psi a)$	$b_2(\phi a . \sim\psi a)$	$b_3(\sim\phi a . \psi a)$	$b_4(\sim\phi a . \sim\psi a)$
$P(b/k_0)$	xy	$x(1-y)$	$y(1-x)$	$(1-x)(1-y)$
$P(b/k_0 . h)$	x_1	0	$y_1 - x_1$	$1 - y_1$

But now how is x_1 related to x, and y_1 to y? The first piece of information taken by itself indicates that x_1 is x, and y_1 is y. The second piece
of information, as we have seen, increases the probability of a future
object being ψ and lowers the probability of a future object being ϕ.
Hence $x_1 < x$ and $y_1 > y$.

(If our information is K_0 that the ratios in the population as a whole
of ϕ's to $\sim\phi$'s are $x:1 - x$ and of ψ's to $\sim\psi$'s $y:1 - y$, then given h, the
proportions of ϕ's which are ψ, ϕ's which are $\sim\psi$, $\sim\phi$'s which are ψ and
$\sim\phi$'s which are $\sim\psi$ in the population will be known for certain. The pro-
portion of $(\phi \cap \psi)$'s will be x, of $(\phi \cap \sim\psi)$'s 0, of $(\sim\phi \cap \psi)$'s $(y - x)$, of
$(\sim\phi \cap \sim\psi)$'s $(1 - y)$, and so, as I argued earlier, the figures in the second
line will be as in Table I.)

Further background evidence about the extent of similarities between
ϕ's and $\sim\phi$'s in their possession of properties like ψ, and between ψ's and
$\sim\psi$'s in their possession of properties like ϕ is going to affect the extent
to which and even whether $x_1 < x$ and $y_1 > y$. If ϕ's are known to have
other properties of a kind similar to ψ or $\sim\psi$ very different from those
possessed by all or most other objects, then y_1 will hardly be greater than
y at all, and could even be less. If h is 'all animals breathe oxygen', we
have a case of this kind. Animals are known to differ markedly from other
objects in their other physiological characteristics. But if ϕ's are known
to have other properties of a kind similar to ψ or $\sim\psi$ very similar to those
possessed by all or most other objects, then y_1 will be significantly $> y$.
If h is 'all objects in Belgium conform to the equations of General Relativity' we have a case of this kind. Objects in Belgium are known not to
differ from other objects (e.g. objects in England) in other physical properties. Similar considerations apply to x_1 and x. If $\sim\psi$'s are known to have
other properties of a kind similar to ϕ or $\sim\phi$ different from those possessed
by all or most ψ's, then x_1 will hardly be less than x and could exceed it.
Whereas if $\sim\psi$'s are known to have other properties of a kind similar to
those possessed by all or most ψ's, then x_1 will be significantly $< x$. In
these inferences background evidence from a wider field (knowledge of
other respects in which ϕ's do or do not resemble $\sim\phi$'s) affects the proba-

bility. I will discuss the role of such background evidence generally and elucidate its influence in Chapter XI. Meanwhile I urge as intuitively obvious that the cited background evidence will have the effects discussed above.

If we suppose, as we have done so far, that the background evidence contains no evidence about the similarities between ϕ's and $\sim\phi$'s, ψ's and $\sim\psi$'s, then, I have argued, we can conclude that x_1 is somewhat less than x, y_1 somewhat greater than y. However, it is more realistic to suppose that knowledge of similarities and dissimilarities between ϕ's and $\sim\phi$'s, ψ's and $\sim\psi$'s found in the past is included in the background evidence. Hence I will suppose in future that k_0 includes such knowledge. This will allow us to consider for what background evidence Nicod's criterion does and does not hold. Normally however it will be the case that the ϕ's mentioned in proposed scientific laws will be, on our background evidence, somewhat similar to $\sim\phi$'s and somewhat different from them in respect of properties similar to ψ; and conversely with ψ's. In which case, as in the case where we do not have such background knowledge, x_1 will be somewhat less than x, y_1 somewhat greater than y.

On this account the difficulty raised by Hooker and Stove [14] can be met. b_3 instances will not usually be significantly confirmatory or disconfirmatory. For y_1 somewhat greater than y, and x_1 somewhat less than x, $(y - xy)$ is liable to be very close in value to $(y_1 - x_1)$. Hence for practical purposes b_3 instances can under these normal circumstances be regarded as irrelevant. However for x_1 only somewhat less than x, given that y is not too close to 1 (which will be the normal case with a scientific hypothesis), b_1 instances will confirm. Also, for y_1 not much greater than y, b_4 instances will confirm – given that x is not close to 0. Even if x is very close to 0, only if y_1 considerably exceeds y will b_4 instances disconfirm; otherwise they will be irrelevant. So in general we will expect b_1 instances to confirm, b_4 instances to confirm or be irrelevant, and b_3 instances to be irrelevant. Further as Mackie and Alexander claim, b_1 instances can in general be expected to confirm much more than b_4 instances, which will lead us to suppose that b_4 instances are irrelevant and so, wrongly, to adopt Nicod's criterion. So for most hypotheses which we are likely to come across, we have all the advantages without the disadvantages of the Mackie-Alexander account.

However, there will be values of x, y, x_1, and y_1, for which b_1 and b_4 instances will be disconfirmatory. None of the detailed solutions proposed so far for the paradoxes allow for this possibility, and the worth of my solution can best be tested by making plausible this surprising consequence. (There will also be values for which b_3 instances will be disconfirmatory. However, we have already shown that this will not happen very often, and that it happens sometimes is less surprising than that b_1 or b_4 instances sometimes disconfirm. If I can justify the latter claim, I will not trouble to justify the less surprising consequence of my account.)

My account shows that 'ϕa. ψa' will disconfirm 'all ϕ's are ψ' if y is very close to 1 and x_1 is significantly less than x. y will be very close to 1 if our background knowledge is that virtually everything in the universe is ψ. x_1 will be significantly less than x if our background knowledge is that ψ's are very similar to $\sim\psi$'s in their properties similar to ϕ. These conditions are satisfied for the following two hypotheses:

h_1: all grasshoppers are located in parts of the world other than Pitcairn Island;

h_2: all monkeys are of height other than exactly 6 feet.

x_1 will be significantly less than x for h_1 since things located on Pitcairn Island are, to our knowledge, much like things located in many other parts of the world, especially in islands of similar climate, in respect of the kind of animal, plant, or inorganic stuff they are. The background evidence suggests that dividing things into two classes, 'located on Pitcairn Island' and 'not located on Pitcairn Island', is not likely to be fruitful for science. y is close to 1 since our background evidence is that a very large proportion of things is not located on Pitcairn Island. Similar considerations show that my conditions are satisfied for h_2. And when we reflect on the effect of a b_1 instance on these hypotheses we realize that in fact it is disconfirmatory. Finding by chance a grasshopper somewhere else than on Pitcairn Island as such (that is, in the absence of further information, e.g. that it was found in a region where grasshoppers were already known to abound) only suggests that grasshoppers are more abundant than we supposed and so in view of the similarities between things located and things not located on Pitcairn Island, more likely than we supposed to be located on Pitcairn Island. We can see the point yet more clearly if we consider the effect on the hypothesis of the discovery of a large number of b_1 instances. Discovery that the rest of the world was swarming with grasshoppers clearly casts grave doubt on the hypothesis. But the discovery of a large number n of grasshoppers can be represented as the discovery of n individual grasshoppers in succession. Either each discovery disconfirms slightly or at some stage there is a sudden large increment of disconfirmation. The latter is implausible, for any choice

of *m,* such that although observation of *m* grasshoppers did not disconfirm, observation of the (*m* + 1)'th grasshopper disconfirmed substantially,
would seem arbitrary. Hence, I conclude, each instance is separately disconfirmatory. Similar considerations apply to h_2. The more monkeys we
find of height other than exactly 6 feet the more monkeys there are
likely to be, and so, in view of the general similarities between things
exactly 6 feet high and other things, the more likely some of the unobserved monkeys are to be exactly 6 feet high.

It may be objected to these examples that they are not of a type which
scientists are likely to put forward. Scientists are unlikely to postulate a
physically necessary connexion between being a grasshopper and being
located in some other place than Pitcairn Island. This is true, for the
reason that just because of the known similarities between the ψ's and
$\sim\psi$'s in question, the hypothesis is unlikely to be true. But the point is
that if a scientist were to put forward the hypothesis, mere accumulation
of b_1 instances is going to disconfirm it, not confirm it. If the world
proves to be swarming with grasshoppers it is less likely that they need a
specialized environment than if there are only very few of them. In the
circumstances described b_4 instances will be of more confirmatory value
(especially b_4 instances acquired in the process of subjecting the hypothesis to test – e.g. taking grasshoppers to Pitcairn Island to see if they
die on arrival. For such procedures see the last few pages of this chapter).

On my account a b_4 instance will disconfirm for *x* close to 0 and y_1
significantly greater than *y*. *x* will be close to 0 if our background knowledge is that virtually nothing in the universe is a ϕ. y_1 will be significantly greater than *y* if our background knowledge is that ϕ's are very
similar to $\sim\phi$'s in their possession of other properties like ψ. These conditions are satisfied for the hypotheses obtained by contraposing h_1 and
h_2. (To contrapose a hypothesis is to write 'all not-*B*'s are not-*A*'s' instead
of the original 'all *A*'s are *B*'.) The conditions are also satisfied for

h_3: all bank employees are subject to acceleration towards the centre
of the Earth in accord with Newton's law of gravitation;

h_4: all men subjected to exceptional radioactivity do not increase in
mass when squeezed.

Clearly the main effect of finding something other than a bank em-
ployee which does not accelerate towards the centre of the Earth in accordance with Newton's law is to suggest that the probability of an object
not being subject to Newton's law is far greater than had previously been
supposed, and so that bank employees not being very different from

other objects in their physical properties are more likely than had been supposed to be subject to a different law of acceleration than h_3.

What I claim to have shown in the last few pages is that b_1 and b_4 instances can be disconfirmatory even when our background evidence is the mere knowledge of the proportions of and similarities between ϕ's and ψ's, some sort of knowledge of which we will normally have. If the background evidence is more extensive, for example if it includes knowledge of other characteristics of the object a examined, b_1 and b_4 instances can easily be disconfirmatory. Will ([15] pp. 54–7) illustrated this with two examples. In one the hypothesis was 'all yellow fever is transmitted by Aëdes Aegypti', a hypothesis once considered seriously by scientists. A disease a was found to be ϕ ('yellow fever') and to be ψ ('have been transmitted by Aëdes Aegypti'). But the background knowledge k included the information that a was situated in an isolated region and that it was extremely unlikely that yellow fever could reach such an isolated region if transmitted only by Aëdes Aegypti. The addition to the background evidence of '$\phi a , \psi a$' therefore cast considerable doubt on h.

The Mackie-Alexander figures in the second line of Table I were challenged in articles by P. R. Wilson [9] and G. Nerlich [10]. Nerlich, developing a point of Wilson's, claimed that the addition of h to the evidence meant that there was a certain amount of probability $x (1 - y)$ previously allocated to a b_2-instance to be distributed among other instances, that there was more than one possible way of doing this, and that Mackie had not shown why his way of distributing it was the correct one. My arguments of the last few pages are an attempt to show what are the correct figures for the second line of the table (and allow for the possibility, which Nerlich did not, that b_1 and b_4 instances may lose probability from the first line to the second).

The quantitative solution, which for the last few pages I have been concerned to refine and justify, was designed to show the falsity and to explain the apparent plausibility of Nicod's criterion. It attempted to show that where it was known that the number of ϕ's was much less than the number of non-ψ's, then '$\phi a . \psi a$' confirmed 'all ϕ's are ψ' much better than did '$\sim\phi a . \sim\psi a$', or '$\sim\phi a . \psi a$', so giving the false impression that '$\sim\phi a . \sim\psi a$' (and perhaps '$\sim\phi a . \psi a$') did not confirm at all. In refining the quantitative solution I have been concerned to show that neither '$\phi a . \psi a$' nor '$\sim\phi a . \sim\psi a$' nor '$\sim\phi a . \psi a$' necessarily confirm 'all ϕ's are ψ' although in general they do. In the refined version, the normal background evidence will be that the number of ϕ's observed is much less than the number of $\sim\psi$'s observed, and for this background evidence the main points made by the original version of the quantitative solution remain. These are, first, that Nicod's criterion is false; and, secondly, that on the background evidence which we often have for scientific hypotheses it will appear to be true, because on this normal evidence '$\phi a . \psi a$' confirms 'all ϕ's are ψ' much better than does '$\sim\phi a . \psi a$' or '$\sim\phi a . \sim\psi a$'. It is however a consequence of this, given the scientific laws condition and the equivalence condition, that, where the number of ϕ's (ϕ's observed in the refined version) is much less than the number of $\sim\psi$'s, '$\phi a . \psi a$' also confirms 'all $\sim\psi$'s are $\sim\phi$' much better than does (e.g.) '$\sim\phi a . \sim\psi a$'. This, Scheffler claims ([8] p. 284) leads to paradoxical results.

To use an example virtually the same as Scheffler's, let us consider the case where 'ϕ' = 'vertebrate', and 'ψ' = 'has kidneys'. The number of (observed) vertebrates is presumably much less than the number of (observed) objects which do not possess kidneys. In that case the discovery of a vertebrate with kidneys ought to confirm 'all non-vertebrates lack kidneys' better than does the discovery of a non-vertebrate without kidneys. But such a result, suggests Scheffler, is paradoxical.

I do not think that this result is at all paradoxical. We must bear in mind that what is discovered is not that a known vertebrate has kidneys, but that a certain object is both a vertebrate and has kidneys. Now we already have good evidence that most things in the universe are non-vertebrates and do not have kidneys. So if something turns up which is non-vertebrate and has no kidneys, this was to be expected whether or not the hypothesis 'all non-vertebrates lack kidneys' is true. Yet the discovery of a vertebrate with kidneys is the discovery that in the small class of objects with kidneys, where alone any exceptions to the hypothesis can be found, one object is not an exception. It therefore seems reasonable to suppose that the latter discovery confirms the hypothesis better than the former one, as the quantitative solution demands.

The apparent paradoxicality of such results as this arises, I believe, from the fact that the $b_1, b_2, b_3,$ and b_4 instances considered are not instances found in the course of pursuing any particular policy (e.g. inspecting ϕ's) but are simply chance observations. But the supposition that we are often in practice concerned with the confirming effect of such chance instances is an unrealistic one. When an hypothesis is up for testing we do not generally note everything we observe, rather we seek to test the hypothesis by making observations of particular kinds.

A different objection to Nicod's criterion from those so far discussed, developed from the point just made, comes in various articles by

J. W. N. Watkins writing from the standpoint of a 'Popperian' philosophy of science. Watkins claimed that whether an observation confirmed an hypothesis is a matter of whether it was made in the process of testing the hypothesis. This would seem to imply that '$\phi a . \psi a$', '$\sim\phi a . \psi a$', and '$\sim\phi a . \sim\psi a$' may each confirm h, 'all ϕ's are ψ', or be irrelevant to it. They confirm only when observed in the process of testing h. If we find an a which is ϕ and ψ when looking for a falsifying instance of h, then, since we found it instead of an a which was ϕ and $\sim\psi$ our observation confirms the hypothesis; but if we had just come across the a which was ϕ and ψ by chance, or, worse still, come across it when setting ourselves to find it, it would not have confirmed the hypothesis. Now this makes the whole matter very psychological. Surely whether or not an observation confirms an hypothesis is not a matter of the intention which the investigator had while making it. Watkins admitted this in [21] and claimed there that an observation b confirmed h if and only if $P(b/h . k) > P(b/k)$. misp This is our Principle C (the main part of Mackie's relevance criterion). How then does the fact of testing come in? Surely in the way suggested by Mackie. If we are looking for ($\phi a . \psi a$), we conduct ourselves so as to be more likely to find it than to find ($\phi a . \sim\psi a$) or ($\sim\phi a . \psi a$) or ($\sim\phi a . \sim\psi a$); for example, we might look only at ψ's. In that case we are not likely to find counter-instances to h if there are any. But this fact is a new piece of background evidence, that we are pursuing a certain policy of investigation and the objects discovered are found in the course of pursuing that policy. This, not the intention of the investigator, is the objective fact to be added to the background evidence, which affects the probabilities of the different discoveries, and so of the hypothesis, given those discoveries. If we are pursuing a policy as a result of pursuing which we are unlikely to find a counter-instance, the confirmation which b_1 gives to h is lessened. The opposite happens if we pursue a policy which means that we are more likely to find a counter-instance if one exists than if we made mere chance observations. This general point can be seen by examples. The point which I make is Mackie's [5], but the figures in my examples differ from his, because of my different assumptions about background evidence and its effect. My figures will be found to bear out the general point better than Mackie's. Some of my results are similar to those of Watkins, although some differ – e.g. Watkins does not allow that a b_1 instance may disconfirm.

Let q_1 be the policy of examining only ψ's to see whether or not they are ϕ's, the hypothesis h being tested, being, as before, 'all ϕ's are ψ'. This policy is such that we will never find a counter-instance to h even if there is one to be found. The probabilities on this policy will therefore be:

TABLE IIa

	b_1	b_2	b_3	b_4
$P(b/k_0 . q_1)$	x	0	$(1-x)$	0
$P(b/k_0 . h . q_1)$	x_1	0	$(1-x_1)$	0

Under these circumstances, b_1 observations disconfirm since $x_1 < x$ unless the background evidence is that ψ's and $\sim\psi$'s are very different in other properties similar to ϕ. 'All ravens are black' is surely not of this latter kind. Hence the discovery of a black raven will be very slight counter-evidence to 'all ravens are black'. This is because by showing that there are more ravens then we had supposed among black objects, it increases the likelihood of there being ravens among non-black objects too. The adoption of q_1 guarantees what in general will not be the case, that (unless ψ's and $\sim\psi$'s are known to be very different from each other in their properties of the type to which ϕ belongs) a b_1 observation is either irrelevant or disconfirmatory. The plausibility of this result can be seen even more clearly by considering hypotheses in which the background evidence is that ψ's are very similar to $\sim\psi$'s in their properties similar to ϕ. Consider:

h_5: all orang-outangs live on Sumatra.

Suppose you examine only objects on Sumatra, and find the place swarming with orang-outangs, viz. many many b_1 instances. This is going to count heavily against h_5, since it suggests that if there are so many orang-outangs on Sumatra there will be at any rate a few to be found elsewhere in view of the known similarities between objects on Sumatra and objects elsewhere in respect of the kind of animal, plant, or inorganic stuff they are. By the argument given earlier, if many b_1 instances disconfirm h massively, each disconfirms it somewhat.

This result, which is surely correct, does not come out of the Alexander-Mackie figures. Table I amended for policy q_1 comes out as follows:

TABLE Ia

	b_1	b_2	b_3	b_4
$P(b/k_0 . q_1)$	x	0	$1-x$	0
$P(b/k_0 . h . q_1)$	x	0	$1-x$	0

On these figures no observation made in pursuit of policy q_1 has any confirmatory or disconfirmatory effect on the hypothesis, which is surely incorrect.

Now let q_2 be the policy of examining only ϕ's for whether or not they are ψ. On my account, the probabilities on this policy will be:

TABLE IIb

	b_1	b_2	b_3	b_4
$P(b/k_0 . q_2)$	y	$1-y$	0	0
$P(b/k_0 . h . q_2)$	1	0	0	0

Hence, given that $y < 1$, necessarily a b_1 instance will confirm. This fact explains our original feeling that b_1 instances must confirm. We were imagining that they were found by looking at ϕ's, the most natural primitive way of testing 'all ϕ's are ψ'.

Serious testing of a scientific hypothesis will normally involve neither policy q_1 nor even policy q_2. It will involve, as Watkins claimed, pursuing some more specific policy q_n which on background evidence alone is very likely to result in finding a b_2 instance if there is one to be found. The pursuit of such a policy will mean that the discovery of a b_1 instance will give more confirmation to the hypothesis than would its discovery if no particular policy was being pursued or policy q_2 was being pursued. For

$\dfrac{P(b_1/k . h . q_n)}{P(b_1/k . q_n)}$ will be greater than $\dfrac{P(b_1/k . h)}{P(b_1/k)}$ or $\dfrac{P(b_1/k . h . q_2)}{P(b_1/k . q_2)}$.

The latter can be seen as follows. Suppose our hypothesis is 'all ravens are black' and our background evidence is that a certain kind of raven is most likely to be non-black, if any ravens are non-black. Then let q_n be q_3, the policy of looking only at ravens of that kind. The pursuit of q_3 instead of q_2 will mean that the figure of the first row of the b_2 column of Table IIb will be greater. The probability of finding a b_2 instance will be, on the background evidence alone, higher if we look among ravens of this kind than if we look at ravens in general. Hence the figure in the first row of the b_1 column will be lower, viz. some value y_2, $y_2 < y$. However the figures in the second row remain the same.

Hence $\dfrac{P(b_1/k . h . q_3)}{P(b_1/k . q_3)} > \dfrac{P(b_1/k . h . q_2)}{P(b_1/k . q_2)}$ since $\dfrac{1}{y_2} > \dfrac{1}{y}$

So the general point made by Watkins is correct. Indeed we have developed it by correcting the figures of Alexander and Mackie.

The arguments of this chapter have confirmed the conclusion of Chapter IV that Nicod's criterion is false. Certainly b_2 instances always disconfirm (subject to the qualifications made at the beginning of the chapter); but whether or not b_1, b_3 and b_4 instances confirm, are irrelevant, or disconfirm is a matter of what our background evidence is and what experimental policy we are pursuing. We cannot suppose a complete absence of background evidence – we must have some knowledge of the frequency and similarities of ϕ's and ψ's – and often when testing an hypothesis we will be collecting observations by pursuing a specific policy. Although we found Nicod's criterion false, we did nevertheless reach the unsurprising conclusion that *normally* 'ϕa. ψa' confirms 'all ϕ's are ψ'. The paradoxes of confirmation are resolved when we abandon Nicod's criterion. This chapter has analysed for a limited kind of background evidence, when 'ϕa. ψa', '$\sim\phi a$. ψa', and '$\sim\phi a$. $\sim\psi a$' confirm 'of physical necessity all ϕ's are ψ', and when each is irrelevant to it or disconfirms it.

One further result of this chapter which leads us marginally to qualify certain earlier results is the result that the addition of 'all ϕ's are ψ' to background evidence which did not contain it makes it somewhat less likely than otherwise that a chance object will be ϕ. If we generalize this result we get the result that when '$Pr(\psi/\phi) = p$' is added to our evidence it becomes somewhat less likely that a chance object will be ϕ if p has a high value and somewhat more likely if p has a low value. The supposition made in earlier chapters that a certain number of ϕ's is equally likely to be observed on any hypothesis of the form '$Pr(\psi/\phi) = p$' must therefore be qualified. This qualification is not however needed if we suppose that the ϕ's are observed not by chance, but as the result of a policy of looking for and examining ϕ's.

Bibliography

For Hempel's exposition of and solution to the paradoxes of confirmation, see:

[1] C. G. HEMPEL 'Studies in the Logic of Confirmation' *Mind*, 1945, **54**, 1–26 and 97–121; reprinted with an additional postscript in C. G. HEMPEL *Aspects of Scientific Explanation*, New York, 1965, pp. 1–51. My page references are to the latter edition.

For the quantitative solution, see:

[2] JANINA HOSIASSON-LINDENBAUM 'On Confirmation' *Journal of Symbolic Logic,* 1940, **5**, 133–48.

[3] D. PEARS 'Hypotheticals' *Analysis,* 1950, **10**, 49–63.

[4] H. G. ALEXANDER 'The Paradoxes of Confirmation' *British Journal for the Philosophy of Science,* 1958, **9**, 227–33.

For Mackie's exposition and development of this, see:

[5] J. L. MACKIE 'The Paradoxes of Confirmation' *ibid.,* 1963, **13**, 265–77; reprinted in P. H. Nidditch (ed.) *The Philosophy of Science* London, 1968, pp. 165–76. My page references are to the latter edition.

For criticism of Hempel's solution and of the quantitative solution, see:

[6] I. J. GOOD 'The Paradox of Confirmation' *British Journal for the Philosophy of Science,* 1960, **11**, 145–9.

[7] I. J. GOOD 'The Paradox of Confirmation (II)' *ibid.,* 1961, **12**, 63–4

[8] ISRAEL SCHEFFLER *The Anatomy of Inquiry,* New York, 1963, pp. 258–95.

[9] P. R. WILSON 'A New Approach to the Confirmation Paradox' *Australasian Journal of Philosophy,* 1964, **42**, 393–401.

[10] G. NERLICH 'Mr Wilson on the Paradox of Confirmation' *ibid.,* 1964, **42**, 401–5.

[11] I. J. GOOD 'The White Shoe is a Red Herring' *British Journal for the Philosophy of Science,* 1967, **17**, 322.

[12] C. G. HEMPEL 'The White Shoe: No Red Herring' *ibid.,* 1967, **18**, 239–40.

[13] I. J. GOOD 'The White Shoe *Qua* Herring is Pink' *ibid.,* 1968, **19**, 156–7.

[14] C. A. HOOKER and D. STOVE 'Relevance and the Ravens' *ibid.,* 1968, **18**, 305–15.

For Will's claim that b_1-instances may disconfirm, see:

[15] F. L. WILL 'Consequences and Confirmation' *Philosophical Review,* 1966, **75**, 34–58.

For the 'Popperian' solution to the paradox and controversy about it, see:

[16] J. W. N. WATKINS 'Between Analytic and Empirical' *Philosophy,* 1957, **33**, 112–31.

[17] C. G. HEMPEL 'Empirical Statements and Falsifiability' *Philosophy*, 1958, **34**, 342-8.

[18] J. W. N. WATKINS 'A Rejoinder to Professor Hempel's Reply' *ibid.*, 1958, **34**, 349-55.

[19] J. AGASSI 'Corroboration Versus Induction' *British Journal for the Philosophy of Science*, 1959, **9**, 311-17.

[20] H. G. ALEXANDER 'The Paradoxes of Confirmation – A Reply to Dr Agassi' *ibid.*, 1959, **10**, 229-34.

[21] J. W. N. WATKINS 'Confirmation Without Background Knowledge' *ibid.*, 1960, **10**, 318-20.

[22] D. STOVE 'Popperian Confirmation and the Paradox of the Ravens' *Australasian Journal of Philosophy*, 1959, **37**, 149-51.

[23] J. W. N. WATKINS 'Mr Stove's Blunders' *ibid.*, 1959, **37**, 240-1.

[24] D. STOVE 'A Reply to Mr Watkins' *ibid.*, 1960, **38**, 51-4.

[25] J. W. N. WATKINS 'Reply to Mr Stove's Reply' *ibid.*, 1960, **38**, 54-8.

[26] J. W. N. WATKINS 'Confirmation, Paradox, and Positivism' in M. Bunge (ed.) *The Critical Approach*, New York, 1964.

CHAPTER XI

Overhypotheses

So far in this book we have discussed and developed some very general principles of epistemic probability, but on the whole applied them only to extremely simple examples. In this chapter I shall consider briefly some slightly more complicated examples of evidence rendering hypotheses probable, and the application to them of the principles developed earlier. When we see the arguments of earlier chapters against the background of the considerations which I shall adduce in this chapter, the examples used there will, I hope, begin to seem more realistic. In this chapter I shall consider one or two of the simpler ways in which nomological propositions, e, affect the probability of other nomological propositions, h.

One possibility is that the nomological propositions of e have a wider scope than does h. They may be the propositions of a very general scientific theory. Extreme and uninteresting cases of this arise where e entails h or where e entails $\sim h$, that is where the hypothesis up for consideration is either entailed by or is incompatible with the general theory, and so for any k $P(h/e \cdot k) = 1$ or $P(h/e \cdot k) = 0$. Cases are possible too where the wider theory gives an intermediate degree of probability to the narrower theory.

More interesting is the case where the evidence consists of nomological propositions $h_2 \ldots h_n$ no wider in scope than the hypothesis up for consideration h_1; yet $h_2 \ldots h_n$ confirm or disconfirm h_1 by confirming or disconfirming overhypotheses, that is theories of wider scope than h_1 which render h_1 probable to different degrees. Goodman has dealt with this kind of case in the last chapter of *Fact, Fiction, and Forecast* [1]. I will begin by setting forward Goodman's account of the matter in his own terminology. According to Goodman, 'a hypothesis is a positive overhypothesis of a second, if the antecedent and consequent of the first are parent predicates of, respectively, the antecedent and consequent of the second.' Recall that (see p. 120) on Goodman's definition, '*A*' is a parent predicate of '*B*' if among the classes to which '*A*' applies is the extension of '*B*'. (In 'all *A*'s are *F*', '*A*' is the antecedent predicate and '*F*' the consequent predicate.) 'Thus if *B* is a small bag [of marbles] and in Utah [and in stack *S*], the hypotheses:

all small bagfuls of marbles are uniform in colour,

all bagfuls of marbles in Utah are uniform in colour,

all bagfuls of marbles in stack *S* are uniformly of same warm colour, and others will be positive overhypotheses of H_4' ([1] p. 111) where H_4 is 'all the marbles in bag *B* are red'.

If an overhypothesis is supported, unviolated, and unexhausted and presumptively projectible (that is, not eliminated by a projection rule – the only one now admitted by Goodman being that cited on p. 104) that reinforces the projectibility of an underhypothesis. That is, in so far as the underhypothesis is supported, unviolated and unexhausted, the status of the overhypothesis means that the positive instances give much greater support to the underhypothesis than they would otherwise do. The greater the support for an overhypothesis, the more it reinforces the projectibility of the underhypothesis. Further, for a given degree of support by instances for it, the more specific the overhypothesis, the more it reinforces the projectibility of the underhypothesis. If *n* bagfuls of marbles from stack *S* in Utah are examined and found to be uniform in colour, this fact supports both 'all bagfuls of marbles in stack *S* are uniform in colour' and 'all bagfuls of marbles in Utah are uniform in colour'. But the fact that it supports the former reinforces the projectibility of H_4, 'all marbles in bag *B* are red', more than does the fact that it supports the latter.

So then the reinforcement of projectibility, that is in my terms confirmation, given to underhypotheses by positive overhypotheses depends on the projectibility, support, and specificity of the overhypotheses. Overhypotheses may also disconfirm hypotheses. If the antecedent of an hypothesis *M* is a parent predicate of the antecedent of an hypothesis *H*, but the consequent of *M* is complementary to the consequent of a parent predicate of *H*, *M* is termed a negative overhypothesis of *H*. (A predicate '*X*' is complementary to a predicate '*Y*' if '*X*' applies to all objects in the universe of discourse to which '*Y*' does not apply, that is if '*X*' is 'not-*Y*'.) Thus 'every bagful in *S* is mixed in colour' is a negative overhypothesis of 'all the marbles in bag *B* are red'. In so far as a negative overhypothesis has projectibility, support, and specificity, that decreases the support given by its instances to an underhypothesis.

Goodman points out the importance of the requirement that the over-hypothesis be projectible:

> If for example, all the naval fleets of the world have been examined and each is found to be uniform in colour, and if the predicate 'bag-fleet' applies to just such fleets and to bagful *B* of marbles, then 'every bagfleet is uniform in colour' is an unviolated, well supported, positive overhypothesis of H_4. Yet obviously our information concerning naval fleets contributes nothing to the projectibility of H_4. ([1] p. 111)

(H_4 is, as before, 'all marbles in bag *B* are red'.)

Now this account of Goodman's needs amendment in one small but important respect. According to Goodman, a projectible positive over-hypothesis *H* has its projectibility reinforced by underhypothesis *h*, *H* being a positive overhypothesis of *h*, if the antecedent predicate of *H* is a parent predicate of the antecedent predicate of *h* and the consequent predicate of *H* is a parent predicate of the consequent predicate of *h*. This however would commit Goodman to the view, to alter his example slightly, that where *B* and *D* are bagfuls of marbles in stack *S* in Utah, that

h_1: all green marbles in Utah are marbles in bag *B*
h_2: all yellow marbles in Utah are marbles in bag *B*
h_3: all red marbles in Utah are marbles in bag *D*

all reinforce the projectibility of the (presumptively projectible) hypothesis

H_1: all groups of marbles in Utah, all members of which have the same colour, are bagfuls in stack *S*.

H_1 entails that each bag in stack *S* consists only of marbles of one colour, an hypothesis clearly falsified by the evidence cited (according to which there are both green and yellow marbles in bag *B*)! To maintain his claim that support for projectible positive overhypotheses reinforces the projectibility of an underhypothesis, Goodman ought to have defined a positive overhypothesis somewhat differently, as follows: *H* is a positive over-hypothesis of *h* if the antecedent of *H* is a parent predicate of the ante-cedent of *h*, and the consequent of *H* is an *S*-parent of the consequent predicate of *h*. I define '*Q*' as an *S*-parent of '*R*' if among the classes to which '*Q*' applies is every class, all of whose members belong to the ex-tension of '*R*'. On this new definition h_1, h_2, and h_3 will reinforce the projectibility *not* of H_1 above (because, 'bagful in stack *S*' does *not* apply to *every* class, all of whose members are marbles in bag D; only to that class which comprises the whole bagful), but of

H_2: all groups of marbles in Utah, all members of which have the same colour, are groups of marbles, all of whose members belong to some one bag in stack S.

Goodman's example is not however spoilt because 'uniform in colour' is an S-parent as well as a parent of 'red'. This is because 'uniform in colour' applies to *every* class of red objects as well as to the class of all red objects.

Having made this slight amendment, let us now put Goodman's points about overhypotheses in our terminology, and then illustrate them with one or two more realistic examples. When it is put into our terminology, Goodman's account can be generalized so as to deal with statistical over-hypotheses (that is, nomological propositions of the form $Pr(Q/A) = p$ where $p \neq 1$ or 0). We are assessing the probability of h_1, 'all A's are Q'. Our background evidence k is a mere tautology. The empirical evidence e has the form of propositions similar to h_1, h_2 'all B's are F', h_3 'all C's are G', and h_4 'all D's are H'. 'Θ' is a parent predicate of 'A', 'B', 'C', and 'D'. 'Ψ' is an S-parent predicate of 'Q', 'F', 'G', and 'H'. The similarity of h_1, h_2, h_3 and h_4 consists in the fact that '$Pr(\Psi/\Theta) = 1$' is a comparatively simple hypothesis. h_2, h_3, and h_4 form our evidence giving '$p_{i-1} < Pr(\Psi/\Theta) \leqslant p_i$' ($p_i$ being as on p. 127), different posterior probabilities for different values of p_i.

h_2, 'all B's are F', says that the class of B's is a class, all of whose members are (of physical necessity) 'F', that is, belong to the extension of 'F'. This entails that there is an object a which is a Θ ('Θ' being a parent predicate of 'B', the class of B's is a Θ) and a Ψ('Ψ' being an S-parent of 'F', *every* class, all of whose members are – of physical necessity – F, is a Ψ). So h_2 entails 'Θa . Ψa'. h_3 entails the existence of another object which is both Θ and Ψ, 'Θb . Ψb'; and h_4 entails the existence of another object which is both Θ and Ψ, 'Θc . Ψc'. If h_2 is 'all the marbles in bag B are red' it entails 'bagful B is uniform in colour'. h_2, h_3, and h_4 thus have the relation of objects which are both Θ and Ψ to hypotheses of the form '$p_i - 1 < Pr(\Psi/\Theta) \leqslant p_i$' described in Chapter VIII. By the considerations of Chapter VIII the more hypotheses there are like h_2, h_3 and h_4, the closer to 1 will they raise the probability of another Θ being Ψ (given that the existence of the new Θ is already known – otherwise we meet the difficulties discussed in Chapter X). In so far as the probability of a bagful of marbles in Utah being uniform in colour approaches 1, and bag B is such a bag, that raises towards 1 the probability of 'either all marbles in bag B are red, or all marbles in bag B are blue, or . . .' and so on through all the colours. It lowers towards 0 other possibilities, such as that some of the marbles in bag B are red and some blue. By lowering the probabilities of other alternatives it raises the probability of 'all marbles in bag B are red', and with the additional information that one such marble is red, raises it towards 1.

From all this Goodman's conclusion follows. Overhypotheses, '$Pr(\Psi/\Theta) = 1$', which are unviolated and unexhausted have their probability raised by instances, that is by underhypotheses; and so (because the probabilities of '$p_i - 1 < Pr(\Psi/\Theta) \leqslant p_i$' for high p_i are also raised – see pp. 133f.) the probability of a certain other Θ being Ψ which is the probability of a certain other underhypothesis being true is raised. However, as we saw on p. 149, we cannot conclude from this that if all bag-fleets (see p. 176) which are naval fleets have been found to be uniform in colour, then the probability of a certain bagfleet (about which it is not known whether it is a naval fleet or bag B) being uniform in colour is close to 1, and we certainly cannot conclude this if the bagfleet is known to be bag B. This is because the evidence about bagfleets can be rephrased in terms of logically equivalent but simpler predicates. (To say that an object is a bagfleet is to say that it is either a naval fleet or bag B. 'Bag B' and 'naval fleet' are simpler predicates than 'bagfleet'.) When this is done no rule of inference which we have stated allows us to make the cited inference. The existence of an overhypothesis '$Pr(\Psi/\Theta) = 1$' for a number of underhypotheses does not raise the probability of a certain new underhypothesis, if 'Θ' is logically equivalent to the disjunction of two simpler predicates '$K \cap \Pi$' and the underhypotheses known so far to be true assert that a K is Ψ, while the new underhypothesis asserts that a Π is Ψ. (Normally overhypotheses do not use predicates which are *logically* equivalent to disjunctions of predicates used by underhypotheses. 'Bag-ful of marbles in Utah' does not *mean* 'either bag B or bag C or bag D' even if in fact bag B, bag C, and bag D are the only bags of marbles in Utah. 'Mammal' does not *mean* 'either cat or dog or . . .' where a finite list of mammals is given; for it is not a logically necessary matter how many genera of mammals there are.)

Further by Principle E (p. 51) on any evidence e an hypothesis H_1 which is entailed by an hypothesis H_2, but does not entail H_2, is no less probable than H_2. Now let both H_1 and H_2 be universal nomological propositions stating that all objects of a certain kind are of physical necessity Θ. (Let H_1 be '$\Pi(\Theta/\Psi) = 1$', and H_2 be '$\Pi(\Theta/\Phi) = 1$'. H_1 being more specific is entailed by but does not entail H_2.) H_1 being more probable than H_2, will, by the weighted sum formula make more contribution to raising the probability of a certain object which is Ψ being Θ, than does H_2 to raising the probability of a certain object which is Φ being Θ. So to the extent to which a more specific overhypothesis is rendered probable by underhypotheses, the more probable is its claim about another underhypothesis. All this is as Goodman claimed.

From our account there follows a more general result than Goodman describes in detail. This is that the more Θ's which are known to be Ψ, that is the more underhypotheses like h_2 are known to be true, and the fewer Θ's which are known to be $\sim\Psi$, that is the fewer underhypotheses like h_2 are known to be false, the greater the probability of another known Θ being Ψ, i.e. of another underhypothesis being true.

Much of the probability conferred by nomological propositions on other nomological propositions can be represented as mediated by over-hypotheses, although propositions sometimes need rewriting in a different way in order for this to be seen. Thus the fact that for many genera of mammals it has been found that all members have four-chambered hearts gives considerable probability to all members of some newly discovered such genus G having four-chambered hearts. Let 'Θ' be 'mammalian genus' and 'Ψ' be 'group, all members of which have four-chambered hearts'. Let A, B, C etc. be animals of different mammalian genera: A be 'cat', B be 'dog', C be 'horse', etc. Let X be 'have four-chambered hearts'. Our evidence consists of h_2 'all A's are X', h_3 'all B's are X', h_4 'all C's are X', and so on. This can be represented as 'Θa . Ψa', 'Θb . Ψb', and 'Θc . Ψc' which, by the principles outlined in Chapter VIII, given 'Θd', confirms 'Ψd', that is that all members of a new mammalian genus will have four-chambered hearts.

This pattern of inference is of course very common in science. The fact that all specimens of many types of acid have a certain property increases the probability that all specimens of another type of acid will have that property; that all specimens of various inert gases behave in a certain way increases the probability that all specimens of another inert gas will also behave in that way. In this chapter we have shown how such inference exemplifies principles of confirmation described earlier in the book.

Bibliography

[1] NELSON GOODMAN *Fact, Fiction, and Forecast*, 2nd edition, Indianapolis, 1965.

CHAPTER XII

Acceptability–I

The main concern of this book has been with the question of when and how much different evidence renders different hypotheses probable. In conclusion it is important to distinguish these questions from two other questions which have in the past often been confused with them. One is the question of when is an inference 'inductively valid'; and the other is the question of when a proposition is 'acceptable', or when does the rational man accept a proposition.

The process of getting from evidence to hypothesis rendered probable by it is naturally regarded as an inference. By a process of argument from the evidence we can reach an hypothesis which is probable or more probable than any rival. Further, the inference is analogous to that from premiss to conclusion in deductive logic. An inference is a deductive one if all it does is to draw out of the premisses propositions already (albeit tacitly) contained in the premisses taken together. So in deductive logic the truth of the premisses makes certain the truth of the conclusion. In the inference from evidence to hypothesis, the truth of the premisses, the evidence, makes probable to some degree the truth of the conclusion, the hypothesis. Certainty is probability of 1. Hence it is natural to regard the inference from evidence to hypotheses when the evidence gives to the hypothesis a probability other than 1 or 0 as an inference similar to but distinct from deduction; and hence such inference has been given the name of induction.[1] Deduction is then regarded as the limiting case of induction. But if we are to regard induction as a species of inference analogous to deduction, it is necessary to make precise what constitutes a valid inductive

(1) Sometimes however the term 'inductive inference' has been confined among inferences of this type to those where the hypothesis has the form 'all A's are B' and the evidence consists of a conjunction of particular propositions reporting of particular A's that they are B. I shall henceforward ignore this narrow sense of 'inductive inference' in favour of the wide sense distinguished in the text.

inference. An inference is deductively valid if the premisses entail the conclusion (that is, if a man who accepted the premisses could deny the conclusion only on pain of self-contradiction). When is an inference 'inductively valid'? Unfortunately there seem to be at least three senses of 'inductively valid' current in writings on this subject. An inference may be regarded as inductively valid

(1) when the evidence renders the conclusion more probable than not, *or*
(2) when the evidence renders the conclusion *h* more probable than any alternative hypothesis, in the sense of any hypothesis incompatible with *h* making equally specific claims about the same objects as *h, or*
(3) when the evidence renders the conclusion acceptable.

Clearly what constitutes a valid inductive inference will vary very much according to which sense we take. (1) may seem the more natural sense. But the step from 'a hundred ravens have been observed to be black' (evidence) to 'all ravens are black', can hardly be described as inductively valid in sense (1). It has however often been put forward as a paradigm of an inductively valid inference. This is surely because 'all ravens are black' is rendered more probable by the evidence than any other hypothesis ascribing a colour to every raven, e.g. 'all ravens are white' or 'all ravens observed so far are black, and all other ravens are white'. In other words 'inductively valid' is understood in sense (2). If the 'inductive validity' of an inference is understood in either of the first two senses, the results of earlier chapters can be applied to determine whether an inference is inductively valid. The 'inductive validity' of an inference in the third sense will depend on how 'acceptable' is understood. The main concern of these last two chapters is with this question of when is a proposition acceptable.

Within the last ten years philosophers have realized that a proposition may be probable on the evidence available, without it being reasonable in some sense of 'accept' to accept the proposition. The evidence may render probable the proposition that a certain drug is harmless without it being reasonable for a physician to accept that proposition in the sense of taking it for granted and so prescribing it to his patients. The evidence would have not merely to render it probable that the drug was harmless but would have to render it very probable indeed before it became reasonable to prescribe it – with qualifications dependent on the effects of not prescribing the drug, the kind of harm to be expected if the drug proved harmful, and so on. The probable is not necessarily the acceptable.

But while it has been a gain that this has been seen, confusion has been generated by the use of the word 'accept'. Talk of scientists 'accepting' propositions is found less commonly among scientists than among philosophers analysing the behaviour of scientists. It has come to be used by both as a technical term, but one used by different writers to correspond to different terms of ordinary language, which it is important to distinguish. '*A* accepts *p*' has been used to mean:

[A]   *A* believes *p*,

[C]   *A* adopts *p* as a claim to be submitted to further tests,

[D]   *A* takes *p* for granted *or A* acts on the assumption that *p*.

(Writers have also used 'reject' as a technical term so that to reject *p* is to accept ~*p*.) Because, for reasons to be given, to believe a proposition *p* is not something which one can decide to do straight off, I shall introduce a further sense of '*A* accepts *p*':

[B]   *A* sets himself to believe *p*.

There has recently been much discussion about whether scientists ever accept theories, but if we distinguish the different uses of 'accept' the answer seems clear. Quite clearly everybody does [A] and [D], scientists also do [C], and some people sometimes do [B].

Now [A] is very different from [B], [C] and [D], for the reason that in this sense of 'accept', one cannot choose or decide to accept a proposition at an instant. At some instant one either does or does not believe it and no choice can affect the matter.

That this is so was a claim of Hume's. 'Belief consists', he wrote, 'merely in a certain feeling or sentiment; in something that depends not on the will, but must arise from certain determinate causes and principles of which we are not masters.' ([1] p. 624). But what Hume did not bring out was that this is a logical matter. It is a logical matter that belief is something that happens to us, not something which we have through choice or decision. For if I did believe some proposition as a result of a decision, I would be aware that this was so. This is because (in the normal or non-Freudian senses of the terms, which I am using) if a man decides to do some action, he knows that he decides to do that action, and if he believes *p*, he knows that he believes *p*. If my belief existed as a result of my decision, I would be aware of my belief coming into existence immediately following my decision to believe and would attribute the belief to the decision as cause. But in that case my belief would not be, and I would know that it was not, something formed by the evidence. Hence I would not have, and would know that I did not have, any grounds for putting any trust in what was believed. But in that case I would not really believe it.

Yet, while a man cannot change his beliefs at will, what he can do is [B], set himself to change them over a period. A man's beliefs depend on his evidence and how he evaluates it, that is sees it as rendering different propositions probable. Hence if you are seeking to change your beliefs, there are two possibilities – to seek new evidence which will render probable different propositions, or to seek to change your criteria for assessing evidence. Now although a man could endeavour to change his beliefs by searching for new evidence, this is not by itself a procedure likely to produce this effect and, if he is rational, the agent will realize this. For if the present evidence e is that p is probable, the evidence is that new evidence will also show p to be probable. A more hopeful method for changing one's beliefs is to set oneself to change one's criteria for assessing evidence, and so to assess the old evidence in a new way. There are various old and new ways of setting about this. One might for example submit oneself to a course of brainwashing from an expert Chinese brainwasher. Setting oneself to change one's criteria for assessing evidence over a period is a logically possible procedure. For even if one is successful, at any *instant* one's beliefs will depend on the evidence then available and the criteria one then has, and be independent of one's immediate choice. However the quicker and more certain of success are one's efforts to change one's criteria of assessing evidence, the closer does the procedure get to the border of the logically impossible. For the quicker and more certain of success were one's efforts, the more it would be the case that one was choosing then and there one's criteria for assessing evidence and so one's beliefs on the available evidence. But if one was choosing one's 'beliefs' one would realize that one was choosing them, and so would realize the unreliability of what was chosen, and so would not really believe them. If one was too successful too quickly one would not believe what on the new criteria for assessing evidence was probable. However with this caution, either method for changing one's beliefs is a logically possible one, and both are methods of doing something. So [B], 'set oneself to believe' like [C], 'adopt as a claim to be submitted to further tests' and [D] 'take for granted' or 'work on the assumption that' are things which one can decide or choose to do.

In this chapter and the next chapter I shall inquire of the different senses of 'accept' when it is reasonable to accept a proposition, what are the grounds for acceptance in each sense of 'accept'. I shall thus elucidate the relation between the probability and the acceptability of a proposition. My general thesis about acceptability in all senses is that, other things being equal, it is reasonable to accept a proposition if it is more probable than any alternative and not otherwise, but that there are also considerations other than probability which make it reasonable or unreasonable to accept propositions. This point will be illustrated in due course for each sense of 'accept'. I shall begin with sense [A]. When will the reasonable man believe a proposition? Although a man cannot immediately help having the beliefs he does, he can be assessed as reasonable or unreasonable in having them.

However, before we can answer this question, it is necessary to answer a preliminary question, what is the relation between probability and actual belief. An obvious suggestion is

[F] A believes $p \equiv$ on the evidence available to A, p is probable (that is, more probable than $\sim p$).

[F] will not however do. A may misassess the evidence, and so believe p, even though p be not probable on the evidence. Such misassessment may take the form of a simple mathematical mistake in applying the rules for assessing evidence. Given evidence that there are six red balls and four green balls in a bag, and that each ball is equally likely to be chosen on any occasion, applying the rules of the probability calculus rather hastily, I may conclude that, if I draw three balls without replacement, it is probable that I will draw two green balls in succession. I would be mistaken. Or my mistake may arise from using wrong rules for assessing evidence. I may believe the innocence of a man accused of murder solely on his own testimony, despite the rival testimony of many witnesses, about whose reliability as witnesses there is much evidence. To meet this difficulty, a second suggestion is

[G] A believes $p \equiv A$ believes that on the evidence available to A p is probable.

This analysis, it is true, defines a belief by a belief, but that does not render it useless, so long as there are independent tests for the truth or falsity of the analysans or analysandum, as there are in this case. We can to some extent find out whether or not a man believes that the earth is flat – e.g. by asking him – without *first* finding out what are his standards for assessing evidence, and conversely.

There is however a more serious difficulty with this analysis, the difficulty concerning what is called deductive cogency. If A believes p, and A believes q, and A believes r, then, we are inclined to say, surely A believes p and q and r. (See Chisholm [2] p. 13 for this claim.) But although on evidence *e p* be probable, and *q* be probable, and *r* be probable, (*p . q . r*) may not be. Let *e* be the evidence of the past behaviour and constitution of the dice, *p* be 'one of the faces 1, 2, 3 or 4 will turn up on the next throw', *q* be 'one of the faces 3, 4, 5 or 6 will turn up on the next throw' and *r* be 'one of faces, 1, 2, 5 or 6 will turn up on the next throw'. Let $P(p/e) = 2/3$, $P(q/e) = 2/3$, $P(r/e) = 2/3$, and let *A* believe these values to hold. Then on analysis [G] *A* will believe *p* and he will believe *q* and he will believe *r*. But $P(p . q . r/e) = 0$ and we can reasonably suppose *A* to believe this. In which case he will certainly not believe *p* and *q* and *r* on analysis [G] or any other remotely plausible analysis. (This difficulty is sometimes called 'the lottery paradox' because of the example by means of which Kyburg – [5] pp. 196ff. – originally introduced the difficulty.)

There are two ways out of this difficulty, neither of them superficially very attractive. One is Kyburg's claim (see [5] pp. 196ff.) that a man may indeed believe *p* and believe *q* and believe *r*, without believing (*p . q . r*). The other is Levi's claim that belief is not related in the kind of way suggested to probability – a man may indeed believe what he believes improbable (see [7] *passim*). I would claim that Levi's point with its connected general approach is correct, and that there follows from it, for certain cases, Kyburg's claim.

The basic claim of Levi's which I shall develop is that belief is relative to alternatives. Hence I propose instead of analysis [G]:

[H] *A* believes $p \equiv A$ believes that on the evidence available to *A p* is more probable than any alternative.

An alternative to a proposition is what it is being contrasted with. The normal alternative to a proposition is its denial, and under these circumstances [H] reduces to [G]. But there are cases when a proposition is regarded as one of several alternatives, at least and at most one of which, given the evidence, must be true. The denial of a proposition is thus in these circumstances broken down into a number of different ways in which the proposition can be false. Clearly by principle E (p. 51) if *p* is more probable than ~*p*, it will be more probable than any proposition *r* which is known to be a member of a disjunction logically equivalent to ~*p*. Hence if a man believes *p* because he believes that *p* is more probable than ~*p*, he will believe *p* whatever the alternatives with which *p* is being contrasted.

A proposition will often be regarded as one of a number of alternatives when it claims that one of a number of possible results (e.g. of a horse race with several horses competing) occurs, or that one of a number of alternative theories, at least and at most one of which must be true, is true. Thus, suppose that, having studied the form for some race, I come to the conclusion that, although it is not probable that Eclipse will win, it is more probable that Eclipse will win than that any other horse will win. I therefore perhaps bet on Eclipse. Suppose now that I am asked 'Whom do you believe will win?' Surely I am not lying if I make the natural answer 'Eclipse' rather than the answer 'I do not believe of any horse that it will win' which is the answer required by [G]. Where there are a number of alternatives (1) . . . (3), all of which may be false, we must, I think, usually add to them the alternative (4), 'None of (1) . . . (3) are true', before sense [H] can have application. Suppose that we are discussing a number of scientific theories, say, in cosmology, and that my judgement of probabilities is that theory (1) is more probable than theories (2) or (3), but that it is still more probable that none of those theories is true. Then it would, I suggest, be more natural to say that I believe none of (1), (2), or (3), than to say that I believe (1). On the other hand if I believe that theory (1) is more probable than the other theories, and that it is more probable that theory (1) is true than that none of them is, then it seems correct to say of me that I believe that theory (1) is true. If A believes p, then also either he will believe that he knows p or he will not have that belief. In the former case he believes that p is among his evidence, in which case he will believe that, given his evidence, p is certain and so has an epistemic probability of 1. Hence A will believe that on his evidence p is probable, and so more probable than any alternative. If A believes p but does not believe that he knows p, then he must believe that he has evidence other than p which makes p more probable than any alternative. He may not be able to state this evidence clearly in words, but he must believe that he has such evidence. A man who claimed to believe that the Earth was flat, but claimed also that he believed that, given what he knew about the world, it was more probable that the Earth was round than that it was flat, would appear to contradict himself. Conversely, as we have seen, if A believes that p is more probable than any alternative, it is correct to say of him that he believes that p. For these reasons I claim that, as stated in [H], to say that one believes that p is to say that one believes that p is more probable than any alternative. Only the context can reveal what counts as an alternative. Generally the only alternative to p is $\sim p$, but sometimes, as in the examples given, there are other and different alternatives.

This point will enable us to solve the difficulty of deductive cogency. Normally to say that A believes p is to say that he believes p to be more probable than $\sim p$; to say that he believes q is to say that he believes q to

be more probable than $\sim q$; to say that he believes r is to say that he believes r to be more probable than $\sim r$. If he believes p and he believes q,
and he believes r, does he then believe $(p.q.r)$? It all depends on what
$(p.q.r)$ is being contrasted with. If it is being contrasted with its denial
$\sim(p.q.r)$, then indeed a man who believes p and believes q and believes
r will not necessarily or in general believe $(p.q.r)$. For if $P(p/e) > P$
$(\sim p/e), P(q/e) > P(\sim q/e), P(r/e) > P(\sim r/e), P(p.q.r/e)$ will not necessarily be greater than $P(\sim(p.q.r)/e)$ either in virtue of the axioms of the
probability calculus or any other reasonable principles of probability,
and a man who understands anything of what he is committed to in his
beliefs will not expect it necessarily to be so. With this understanding of
the alternative to $(p.q.r)$ Kyburg is correct. A man may believe p, and
believe q, and believe r, without believing $(p.q.r)$. Clearly this is the
case for the dice example considered earlier. The agent clearly does not
in that example believe $(p.q.r)$ if he is anything like rational. Yet surely
he does believe each of p and q and r; if a betting man, he would be prepared to bet on each at evens. Kyburg's solution only seems implausible
if we have something like a dispositional theory of belief and suppose
that if A believes p, then he must use p as a premiss in his practical inferences. For in that case if p is a premiss of practical inferences and so
is q, and thus we take p for granted and q for granted in those inferences,
surely we must take $(p.q)$ for granted. But this is a mistaken type of
analysis. A doctor may, to use my earlier example, believe that a certain
drug is harmless, without being sufficiently convinced to prescribe it for
his patients. Once one realizes that belief is not so simply connected with
action, Kyburg's solution, I would suggest, loses its initial implausibility.

However, the alternatives with which $(p.q.r)$ is being contrasted may
not be merely $\sim(p.q.r)$, and, according to what the alternatives are, so
the meaning of the claim that A believes $(p.q.r)$ differs. One of the most
common cases is where $(p.q.r)$ is being contrasted with other conjunc-
tions of three propositions, of which either p or $\sim p$, q or $\sim q$, r or $\sim r$ are
members; i.e. it is being contrasted with $(\sim p.q.\sim r), (p.\sim q.\sim r)$, and so
on. Such alternatives are more normal where p, q, and r are believed by A
to be independent on his evidence e; that is, where A believes that
$$P(p/q.r.e) = P(p/e) = P(p/q.e)$$
and the similar claims obtainable by interchanging 'p', 'q', and 'r' in this
equation. If A's beliefs are correct and if $P(p/e) > P(\sim p/e), P(q/e) >$
$P(\sim q/e)$, and $P(r/e) > P(\sim r/e)$ then, it is provable from the axioms of the
probability calculus, $P(p.q.r/e)$ is greater than the probability on evidence e of any other conjunction of three propositions, of which p or

N

$\sim p$ is one member, q or $\sim q$ another, and r or $\sim r$ the third. Hence I suggest that if:

(1) a man believes a number of propositions to be independent on his evidence; and he derives the right consequences therefrom in accordance with the axioms of the calculus about the probabilities of conjunctions,

(2) in saying that he believes the conjunction of those propositions, we mean that he believes it to be more probable than any other conjunction of the same number of propositions to which every member of the original conjunction or its denial belongs,

(3) in saying of each of the conjuncts that he believes it, we mean that he believes it to be more probable than its denial,

then necessarily if a man believes each member of a conjunction, he believes the conjunction.

Thus, to take an example of a conjunction with only two members, suppose a man A believes p, 'Oxford will win the boat race', in the sense that he believes that it is more probable that p than that $\sim p$. Suppose A also believes q, 'Tal will win the world chess championship', in the sense that he believes that it is more probable that q than that $\sim q$. Suppose also that he believes p and q to be independent on his evidence e, as he normally would do in this case, and that he derives the right consequences from all these beliefs in accordance with the axioms of the calculus, that $(p.q)$ is more probable than $(p.\sim q)$, $(\sim p.q)$, or $(\sim p.\sim q)$. Then if we are contrasting $(p.q)$ with the alternatives $(p.\sim q)$, $(\sim p.q)$ and $(\sim p.\sim q)$, we can say that A believes $(p.q)$.

Having discussed the relation between probability and actual belief, we are now in a position to face the question of when the reasonable or rational man will believe a proposition, or when a proposition is worthy of belief.

It is natural to suppose that a proposition which a rational or reasonable man would believe is itself a rational belief; and that a proposition which is worthy of belief is itself credible. Corresponding to the analysis of actual belief [H] there is an obvious analysis of a rational belief:

[I] p is a rational belief for A $\equiv$ on the evidence available to A, p is more probable than any alternative.

The rational belief is that which is based on the evidence, whether or not the believer assesses the evidence correctly. [I] is therefore, I suggest, a correct account of the relation of rational belief to probability. But in this case, to say that a belief is rational or credible is not necessarily, despite etymology, to say that it is worthy of belief or 'good to be believed'.

For, as Firth [3] urged against Chisholm ([2] p. 5) who suggested such an equivalence, saying that it would be good if *p* was believed is making an evaluative claim not made by saying that something is probable or more probable than something else.[1] For there are reasons why it is good if people believe things other than the reason that the evidence supports these things. It would be a good thing . . . for her peace of mind, for the happiness of the children, etc., if Jones's wife believed him faithful (when the only alternative is that he is not faithful), even though the evidence which she has does not on balance render probable the claim that he is faithful. Yet it seems proper to describe this situation as one where it is not a rational belief for his wife that Jones is faithful, or one where Jones's fidelity is not credible. So to say that a proposition is credible, or that a belief is rational, is not to say that it is worthy of belief; it is only to say that it is worthy of belief in so far as the only relevant consideration is rational assessment of the evidence. So a rational belief is not the same as a proposition which ought to be believed. But if a proposition *p* ought to be believed, then *p* is a proposition which it is reasonable to believe, and which a reasonable man will believe, and perhaps one which a rational man will believe. Hence we must distinguish a rational belief from a belief which a reasonable man will hold. Our original question concerned the latter; we were asking of each of the different senses of 'accept' when it is reasonable to 'accept' propositions. [I] provides an analysis of rational belief but not, for the reasons given, of a belief which it is reasonable to hold.

However, although it may be reasonable on occasion to hold a belief which is not more probable than any alternative, I would suggest that, *other things being equal,* it is the reasonable thing to believe a proposition if it is more probable than any alternative, and otherwise not to believe it. For what a man believes, he believes, we have seen, to be more probable than any alternative. A man's beliefs about what is probable determine his actions. If *A* believes that Eclipse will probably win the race, then, if the odds are favourable and he is a betting man, he will bet on it. But only if his judgements of probability are not far wrong is it in general probable that his actions will achieve their intended end. If, despite *A*'s beliefs, it is not probable that Eclipse will win the race, then it is not probable that *A* will win his bet. Since a reasonable man will take steps to achieve what he wants or thinks that he ought to achieve, he will be reasonable to believe a proposition in so far as it is probable, other things being equal.

(1) Chisholm seems subsequently to have accepted this objection. See Roderick M. Chisholm *Theory of Knowledge,* Englewood Cliffs, N.J., 1966, pp. 12ff.

But as we have seen earlier, there are other reasons for belief, e.g. to attain peace of mind. So the probability of a belief, the belief being a rational belief, is not the only thing which makes it reasonable or unreasonable to have it.

Similar considerations obviously apply to sense [B], 'set oneself to believe'. Since, other things being equal, it is reasonable to believe a proposition in so far as it is more probable than any alternative, then, other things being equal, it is reasonable to set yourself to believe what is more probable than any alternative. Hence men ought normally to choose their criteria for weighing evidence so that they come to hold beliefs which are indeed more probable than any alternative, although, as we have seen, there could be grounds making it reasonable for a man to set himself to have beliefs which are not more probable than any alternative.

The relation of 'rational belief' to probability has been the concern of a number of writings of recent years which we must now assess.[1] The most influential of these has been Isaac Levi's *Gambling with Truth* [7]. Levi is concerned with the conditions for 'rational acceptance' of propositions, and by 'acceptance' he means 'belief' ([7] p. 25). Levi shares the view put forward earlier that whether p is a rational belief depends on what are seen as the alternatives to p. In any situation of problem solving facing an investigator, and evidence e there are, according to Levi, a number of 'relevant answers' M_e, which will consist of the members of an ultimate partition U_e, the conjunction of members thereof, and all the different possible disjunctions thereof. The members of the ultimate partition are sentences (Levi prefers to talk of sentences rather than propositions. I ignore the awkwardness of this, referred to on p. 2) of which the evidence (background evidence b plus new evidence e) entails that at least and at most one is true, and which constitute the strongest (that is, the most informative) relevant answers. Thus, to take Levi's example ([7] p. 36), suppose an election forecaster is interested in predicting which of the three candidates X, Y, and Z, will win some election. He is not interested in who will come second, or who will win some other election. Then his ultimate partition 'can be represented by the three sentences' 'X will win', 'Y will win', 'Z will win'. The set M_e (of relevant answers) generated by this ultimate partition consists of the eight sentences listed below:

(i) either X or Y or Z will win (S)

(ii) either X or Y will win

(1) See for example, as well as those to be discussed, [4] and the collection [11].

(iii) either X or Z will win
(iv) either Y or Z will win
(v) X will win
(vi) Y will win
(vii) Z will win
(viii) X and Y and Z will win (C) ([7] p. 36)

To accept C is to contradict oneself. To accept S is to make no prediction beyond the evidence. To accept any sentence involves accepting any
sentence entailed by it – thus accepting (v) involves accepting (i), (ii) and
(iii). Levi's question is which is the strongest sentence, that is the most
informative sentence, the sentence with greatest content, which the cognitive investigator ought to accept.

The cognitive investigator is for Levi the man concerned not with
practical decisions but with finding things out. Levi sees him as having
two goals. One is that he accepts only what is true. 'However, were this
the only desideratum involved, he would obtain a satisfactory result with
no risk whatsoever' ([7] p. 57) by accepting (i) as 'the strongest sentence
accepted as true via induction from his total evidence b & e'. But given
that 'investigators are sometimes rationally warranted in risking error' it
follows that some other desideratum is involved, and that is 'relief from
agnostic doubt' ([7] p. 58). An investigator risks error in order to hold
an informative belief. Relief from agnosticism is to be measured by the
content of a proposition; the more a proposition says, the more relief
from agnosticism is provided by accepting it. Accepting H as the strongest
sentence, given evidence e, provides 'a degree of relief from agnosticism
equal to cont (H, e), which equals m/n where n is the number of elements
in U_e and m is the number of elements in U_e that are inconsistent with H'
([7] p. 74). Thus in the example cited earlier, accepting (i) would provide
0 relief from agnosticism (since no elements of U_e are inconsistent with
it); accepting (ii), (iii) or (iv) would provide 1/3 relief from agnosticism;
accepting (v), (vi) or (vii) would provide 2/3 relief from agnosticism; accepting (viii) would provide relief from agnosticism of degree 1.

Clearly we ought to accept propositions which provide high relief
from agnosticism only if there is some likelihood of their being true. Levi
suggests a rule for balancing the two requirements of truth and relief from
agnosticism. His Rule A is the core of his system.

Rule A: (a) Accept b & e and all its deductive consequences;
        (b) Reject all elements a_i of U_e, such that $P(a_i/e) < (q$ cont-
           $(\sim a_i/e))$ – i.e. accept the disjunction of all unrejected ele-
           ments of U_e as the strongest element in M_e accepted via
           induction from b & e;

(c) Conjoin the sentence accepted as strongest via induction according to (b) with the total evidence b & e and accept all deductive consequences;

(d) Do not accept (relative to b, e, U_e, the probability distribution, and q) any sentences other than these in your language. ([7] p. 86)

q is an index of 'degree of caution' ([7] p. 88), showing the respective weight to be given to the two desiderata, truth and relief from agnosticism. If q = 0, the investigator will accept only b & e and its deductive consequences, and so is a sceptic. 'As q increases, more elements of U_e will, in general, be rejected, thereby [in virtue of (b)] leading to the accep tance of stronger elements of M_e as strongest via induction. In some cases . . . an increase in q will not lead to this result, because of the features of the probability distribution' ([7] p. 88). $0 \leqslant q \leqslant 1$.

In the example cited on pp. 190f. if each element of the ultimate parti tion had the same probability, then only (i) could be accepted, whatever the value of q. This application of the rule brings out the fact that, intuitively, there is no reason for accepting any one of the strongest senten ces (v), (vi) or (vii) rather than any other one, and no reason for accepting any one of the weaker sentences (ii), (iii) or (iv) rather than any other one. So the investigator has to fall back on (i). In no circumstances, by Levi's rule, will the investigator ever accept a self-contradictory sentence such as (viii). If the probability of one element of the ultimate partition, say 'X will win', were 1/2, and of the others 1/4, an investigator with q index > 3/4 would accept 'X will win' as strongest. It is compatible with following Levi's rule that one may accept propositions with probabilities ≤ 1/2 or reject those with probabilities > 1/2. Whether one accepts a proposition or not depends on one's q-index and on the distribution of probabilities among the elements of the ultimate partition.

Now if, as Levi writes, our twin aims in 'accepting' propositions are truth and relief from agnosticism, and if, as Levi assumes, to accept two propositions is to accept their conjunction, certain requirements follow for any two incompatible propositions p and r. We must not accept both of them, for their conjunction is a certain falsehood. If p is more probable than r, and has greater content than or the same content as r, then, if we accept either of them, we ought to accept p. If p and r are equally probable, but p has greater content than r, then, if we accept either of them, we ought to accept p. But it is compatible with the aims to reject both p and r in these two cases. Further, if p is more probable than r, but r has greater content than p, there are various ways in which the weighing of these two factors against each other could be done, some of which would lead us to accept *p,* and some of which would lead us to accept *r.* Levi's Rule A, for any choice of *q,* is only one such possible rule of comparison.

An alternative rule compatible with the above aims, can be produced by amending (b) in Levi's Rule A, to read as follows:

(b) Reject all elements a_i of U_e such that $P(a_i/e) < q$ cont $(\sim a_i/e)$, and then choose freely one of the remaining elements of U_e as the strongest element in M_e accepted via induction from b & e.

Levi's only comment on such a suggestion is that 'in the case of cognitive decision problems of the sort under consideration here, free choice does not seem legitimate' ([7] p. 84), but he does not say why. Further, the choice of *q* seems an arbitrary matter. Adoption of any *q* value, and so acceptance of any of many different strongest sentences is, according to Levi, compatible with rationality. Another criticism is that different rules from Levi's could be given for measuring content, without being in any obvious way less rational than his.

One of the peculiar features of Levi's rule has been pointed out by Lehrer [8]. Lehrer pointed out that often on Levi's rule even a rational man of minimum caution ($q = 1$) could not accept a proposition which was a member of the ultimate partition and whose probability was greater than 1/2, and considerably exceeded that of any other member of the ultimate partition. Thus, if the probability on the total evidence (b & e), of *s* is 1/12, of *t* is 1/3, and of *u* is 7/12, where *s, t* and *u* are the three members of the ultimate partition, even for $q = 1$, only ($t \vee u$) can be accepted, not *u.*

Lehrer felt that the fact that the probability of *u* considerably exceeded that of its nearest rival *t* was good reason for accepting *u* and therefore he proposed substituting for (b) in Levi's Rule A

(b_R) Accept as strongest **via** induction from *e* the member, or, if more than one, the disjunction of all those members, a_i of U_e such that for any other member of U_e, a_j, if $p(a_i/e) \neq p(a_j/e)$ then $p(a_i/e) - p(a_j/e)$ is at least *r.* ([8] p. 293)

r serves, as *q* served, as an index of caution. The more cautious man will have a large *r.* Lehrer's rule states in effect that we can adopt an hypothesis if its probability exceeds that of its nearest rival by a sufficient amount. Like Levi's rule however, there seems an essential arbitrariness about Lehrer's rule. Why should difference in probability values be so important, and why adopt one r value rather than any other? Adoption of any r value, and so acceptance of many different strongest sentences is, according to Lehrer, compatible with rationality.

In a later article [9] Lehrer proposed an acceptance rule far less arbitrary than that proposed in [8], in having in it no arbitrary constant such as r. [1] He supposes that there are a number of hypotheses $h_1 \ldots h_n$ providing answers to some question, which are mutually exclusive and exhaustive (that is, in Levi's terms, are members of an ultimate partition). Other hypotheses formed by logical operation on these may also be expressed in the language, e.g. $\sim h_1$ or (h_5 v h_6). He defines the 'direct inductibility' of k from evidence e - $D(k, e)$ - (which seems - see later - to mean the credibility in some sense of k, given e) as follows, by his rule RDI:

RDI. $D(k, e)$ if it is not the case that $e\vdash\sim k$, and for any h (where h is any hypothesis expressible in the language), if it is not the case that e & $k\vdash h$ then $P(k/e) > P(h/e)$. ([9] p. 107) ('$e\vdash k$' means that e entails k.)

Thus if there are four hypotheses h_1, h_2, h_3 and h_4 and $P(h_1/e) = 0.3$, $P(h_2/e) = 0.3$, $P(h_3/e) = 0.3$ and $P(h_4/e) = 0.1$, we can 'directly induce' $\sim h_4$ from the evidence.

Once a proposition has been directly induced from evidence, it can then in Lehrer's system be added to the evidence, and be used to derive any further hypotheses 'inducible' by RDI. However in the example just adduced the addition of $\sim h_4$ to the evidence will allow us to derive nothing further. $\sim h_1$, $\sim h_2$ and $\sim h_3$ have each a probability of 0.7 (on the old evidence, 0.66 on the new) and there is no more probable hypothesis which can be expressed in the language. However if h_1, h_2, h_3, h_4 each had different probabilities, then repeated applications of RDI would eliminate the least probable at each stage, eventually allowing us to induce the most probable.

But Lehrer's rule seems peculiar. Suppose there are four hypotheses, h_1, h_2, h_3 and h_4. $P(h_1/e) = 0.4$, $P(h_2/e) = 0.3$, $P(h_3/e) = 0.2$, $P(h_4/e) = 0.1$. We can then by repeated applications of RDI induce h_1. But now suppose that the hypothesis h_1 has the same probability as before, while the other hypotheses have the same probability as each other: $P(h_1/e) = 0.4$, $P(h_2/e) = 0.2$, $P(h_3/e) = 0.2$, $P(h_4/e) = 0.2$. Now nothing can be induced. $P(\sim h_2/e)$, $P(\sim h_3/e)$ and $P(\sim h_4/e)$ having the same value, we cannot get started in the inductive process. But surely h_1, having the same proba-

(1) Lehrer has proposed this rule and developed it in an interesting way also in [10] and in an article 'Justification, Explanation, and Induction' in [11].

bility in both cases, is, if anything, more acceptable if it exceeds its nearest rival by a greater amount.

This whole discussion of the theories of rational acceptance of Levi and Lehrer has been undertaken because of the considerable prominence which they have achieved in the literature of the subject and the illuminating thoroughness with which the ideas have been worked out. But the impression which they leave with this writer, as has been stressed, is one of great arbitrariness. Why adopt any one such system rather than any other? To answer that question, we must ask ourselves what do these writers mean by 'accept' and what do they mean by 'rational'. Levi, as we have noted, means by 'accept' 'believe'. Although Lehrer does not state in [8] what he understands by 'accept', yet since he writes there in detailed commentary on Levi, I assume that there he also understands by 'accept' 'believe'. But what do Levi and Lehrer mean by a 'rational belief' – what I have analysed as a rational belief, that is a belief which is more probable on the evidence than any similar alternative; or a belief which it is reasonable for a man to hold? If they mean by rational belief what I mean by rational belief, their accounts of it obviously fail. A belief is a rational one if it is more probable on the evidence than any similar alternative, and content and indices of caution are irrelevant.

The alternative is that Levi and Lehrer (in [8]) are concerned with which belief it is reasonable to hold; this is the most natural interpretation. Yet Levi emphasizes that he is concerned only with the 'cognitive investigator'. Levi regards the cognitive investigator as concerned to attain truth and relief from agnosticism, and he and Lehrer [8] provide various ways for the cognitive investigator to have beliefs in which these factors are given different degrees of importance. However it is a consequence of their systems that there are some propositions which are 'rational beliefs' for a man, but which he can only believe if he holds false beliefs about their probability, and propositions which are not 'rational beliefs' for a man, but which he can only not believe if he holds false beliefs about their probability.

This comes about as follows. We saw earlier that A believes p if and only if he believes p to be more probable than any similar alternative. Now if a different understanding of 'rational belief' from my understanding of it be adopted, as it is by Levi and Lehrer [8], there will be propositions which are 'rational beliefs' although they are not more probable than any similar alternative, or propositions which are not 'rational beliefs' although they are more probable than any similar alternative. In respect of these propositions a man could only have a 'rational' belief or disbelief about them, if he had a false belief about their probability. One can understand that there could be compelling practical reasons of the kind described earlier (e.g. to secure peace of mind) which make it reasonable for a man to believe a proposition p and so to believe p to be more probable than any alternative, when in fact (on the evidence available) p is not more probable than any alternative, and conversely; but Levi and Lehrer would seem to have very odd standards if they commend the *cognitive investigator,* the man concerned with finding things out, to hold false beliefs about the probability of propositions.

In [9] Lehrer states that he is concerned to produce a rule allowing us to 'induce' h from e. But this is clouding the issue with a most unhelpful technical term, since, as we have seen (p. 181), it is very unclear what is meant by an 'inductively valid' inference.

Yet since Lehrer follows in Levi's tradition, it seems reasonable to suppose that he means by 'h is inducible from e' 'e renders h acceptable' in the sense of 'credible' or 'worthy of belief' and so that he means either that h is a rational belief given e, or that it is reasonable for a cognitive investigator to believe h, given e. Hence similar comments on [9] to those made about [7] and [8] are appropriate. If Lehrer means by 'h is inducible from e' 'h is a rational belief given e', his system for assessing inducibility is false. For a proposition h could be more probable on evidence e than any similar alternative without it being able to be induced from e by Lehrer's rules (as in the example which I put forward on pp. 194f.). And if Lehrer is concerned with what is a reasonable belief for the cognitive investigator, he is commending him to hold false beliefs about probabilities. For where, as on pp. 194f., there are four (exclusive and exhaustive) hypotheses h_1, h_2, h_3 and h_4 and $P(h_1/e) = 0\cdot4$, $P(h_2/e) = 0\cdot2$, $P(h_3/e) = 0\cdot2$, and $P(h_4/e) = 0\cdot2$, since we cannot induce h_1 from e, by Lehrer's rules, Lehrer is commending us not to believe h_1 and so not to believe h_1 to be more probable than any similar alternative. But h_1 is more probable than any similar alternative. So Lehrer is commending the cognitive investigator to hold false beliefs about probabilities! If Lehrer has some other understanding of 'inducible' he must state it, before we can assess the worth of his system.[1]

(1) Similar difficulties arise with the interesting article [6], one of the co-authors of which is Lehrer. The authors discuss the question of whether it is 'reasonable' to 'accept' a proposition p, when it is agreed the evidence renders p and $\sim p$ equally probable. Since the authors discount as irrelevant the practical advantages of accepting some proposition, and say (p. 46) that they are concerned with 'canons of epistemic rationality', they seem to be concerned with rational belief in our sense. If this is so, the answer seems clear that belief that p is not in the circumstances cited rational belief.

Bibliography

[1] DAVID HUME *A Treatise on Human Nature* (first published 1739), ed. by L. A. Selby-Bigge, Oxford, 1888, Appendix pp. 623-7.

[2] R. M. CHISHOLM *Perceiving,* Ithaca, New York, 1957, Chapters 1 and 2.

[3] RODERICK FIRTH 'Chisholm and the Ethics of Belief' *Philosophical Review,* 1959, **68**, 493-506.

[4] JAAKO HINTIKKA and RISTO HILPINEN 'Knowledge, Acceptance, and Inductive Logic' in Jaako Hintikka and Patrick Suppes (eds.) *Aspects of Inductive Logic,* Amsterdam, 1966.

[5] HENRY E. KYBURG *Probability and the Logic of Rational Belief,* Middleton, Connecticut, 1961, pp. 196-9.

[6] KEITH LEHRER, RICHARD ROELOFS and MARSHALL SWAIN 'Reason and Evidence: An Unsolved Problem' *Ratio,* 1967, **9**, 38-48.

[7] ISAAC LEVI *Gambling with Truth,* New York and London, 1967.

[8] KEITH LEHRER 'Induction: A Consistent Gamble' *Nous,* 1969, **3**, 285-97.

[9] KEITH LEHRER 'Induction, Reason, and Consistency' *British Journal for the Philosophy of Science,* 1970, **21**, 103-14.

[10] KEITH LEHRER 'Theoretical Terms and Inductive Inference' in Nicholas Rescher (ed.) *Studies in The Philosophy of Science* (American Philosophical Quarterly Monograph No. 3), Oxford, 1969.

A recent collection devoted to the issues discussed in this chapter is:

[11] MARSHALL SWAIN (ed.) *Induction, Acceptance, and Rational Belief,* Dordrecht, Holland, 1970.

CHAPTER XIII

Acceptability–II

Having dealt in the last chapter with the acceptability of propositions in senses [A] and [B], we pass on in this chapter to consider acceptability in senses [C] and [D].

Let us begin with sense [D]. In sense [D] to 'accept' a proposition is to take it for granted, or to act on the assumption that it is true. We take various propositions for granted in the course of our theoretical investigations into the truth of other propositions. For example, we may in the course of scientific investigations take for granted various theories about how scientific instruments work, or other auxiliary scientific theories. We can work with such theories, that is 'accept' them in sense [D] without actually believing them. And in the course of making practical decisions we may take for granted various propositions, e.g. that there is enough petrol in the tank or food in the larder, without necessarily believing them true. As with senses [A] and [B], so with sense [D], there are considerations other than their probability which make it reasonable to 'accept' propositions. Thus, other things being equal, one reason for taking an auxiliary theory for granted in theoretical investigation is that it is easy to handle, and another reason is that it is difficult or expensive to test. And it is reasonable to take a proposition for granted in making practical decisions if one can achieve some aim only if that proposition is true. It is reasonable to take for granted that there is a way out of an underground cavern in which one is trapped, for, if there is not, life and freedom cannot be had. However, other things being equal, it is surely reasonable to 'accept' propositions in sense [D] if they are more probable than alternatives, and not otherwise. Thus in one's theoretical investigations one is seeking theories which are probably true (or have high verisimilitude – see later in the chapter). In one's investigations one has to take for granted various auxiliary theories, e.g. pieces of mathematics and theories about the working of instruments. The lower is the probability of such auxiliary theories, the more probable it is that one will reach false conclusions in one's theoretical investigations – one will conclude that a falsified theory has not been falsified, or that a theory which has not been falsified has been. Hence in so far as one takes for granted improbable auxiliary theories one's theoretical investigations are less likely than otherwise to provide a correct answer as to which of the main theories being examined is true. In one's practical decisions one is seeking to satisfy one's aims. Yet in so far as the propositions taken for granted in making practical decisions are unlikely to be true, the conclusions about how to satisfy one's aims are less likely than otherwise to be true. Hence in sense [D] of 'accept', at any rate on plausible assumptions about the point of accepting propositions, other things being equal, it is reasonable to accept those propositions which are more probable than alternatives.

However other factors than probability are relevant. How probable a proposition has to be for it to be acceptable will depend on the relative disadvantages of not accepting it if it is true and of accepting it if it is false. The greater are the latter disadvantages in relation to the former the greater the probability which a proposition will need to have in order to make it acceptable. This idea lies behind theories of 'testing hypotheses' current in mathematical statistics. (For commentary on such theories see [2] Chapters 6 and 7.) Statisticians talk of 'accepting' and 'rejecting' hypotheses and they seem usually to be using 'accept' in sense [D] and 'reject' in the sense that to reject p is to accept $\sim p$ in sense [D]. In their terminology to test an hypothesis h is to do some experiment with a rule that we accept h if the experiment has a certain outcome and reject h if it has a certain other outcome (and perhaps do neither if there is a third outcome). The best known statisticians' theory of testing is that of Neyman and Pearson. They recommend that we should use on any hypothesis h tests of 'low size', that is tests such that the probability of their rejecting h if h is true is small, and of 'high power', that is tests such that the probability of their rejecting h if h is false is high. If the only alternatives are to accept or not to accept, this amounts to recommending that we use tests such that the probability of their not accepting h if h is true is small and of their not accepting h if h is false is high. However Neyman and Pearson admit that how we are to balance size against power depends on the relative disadvantages of accepting h if h is false and rejecting h if h is true. 'The use of these statistical tools in any given case, in determining how the balance should be struck, must be left to the investigator' ([1] p. 296).

Sense [C] of 'accept' is perhaps not a very natural understanding of 'accept', but it is that introduced into the literature of the philosophy of science by Popper. We 'accept' a theory, Popper writes, 'in the sense that we select it as worthy to be subjected to further criticism, and to the severest tests we can design' ([3] p. 419). Popper has put forward a detailed account of the grounds for subjecting theories to test which has been highly influential, and which will for that reason receive in this chapter detailed consideration. The propositions which we 'accept' in this sense may of course be of various kinds, and my final conclusion about the grounds for acceptance will apply to propositions of all kinds. However since Popper was concerned chiefly with the grounds for acceptance of scientific theories, that is bodies of nomological propositions, I will in discussion of Popper's account confine myself to the grounds for accepting nomological propositions.

Karl Popper's account was originally published in *Logik der Forschung* in 1934, an English edition of which with new appendices appeared as *The Logic of Scientific Discovery* [3] in 1959. I will begin by setting forward the account given in this work of the grounds for 'acceptance' in Popper's sense [C].

According to Popper, scientific theories are to be 'accepted' in so far as they are falsifiable (that is, testable) and in so far as they have survived severe tests. Popper's claim has often been misunderstood. He does not claim that a scientific theory which satisfies his criteria is for that reason at all likely to be true. As stated, for Popper to 'accept' a theory is not to judge it true or probable or more probable than a rival theory, but simply 'to select it as worthy to be subjected to further criticism and to the severest tests we can design' ([3] p. 419). Hence a conclusion of his that we ought to 'accept' theories which are 'improbable' is not in any obvious way paradoxical. Any remaining ghost of paradoxicality disappears when we realize that for Popper to say that something is 'probable' is not to say that it is likely to be true. Probability means, for Popper, 'logical probability', and 'logical probability' is simply a measure of lack of content (see [3] p. 119). A statement has smaller 'logical probability' in Popper's sense, the larger the claim it makes about the world. 'Logical probability' in Popper's sense is not to be confused with probability as understood by the Logical Theory of probability, that is, epistemic probability. So then, according to Popper, a theory is to be accepted in so far as it has great testability and has been well corroborated. A theory will have great testability (that is, falsifiability) in so far as there are many opportunities for it to be refuted by experience (that is, to be shown false if it is false). Having obtained a theory of great testability, we proceed to test it. We try to show it false. In so far as sincere tests of the theory fail to show it false, the theory is corroborated. Among highly testable theories we accept, at this second stage, theories which are best corroborated. In so far as a theory can be exposed to more tests, so it can be better corroborated. Testability may therefore be equated with corroborability.

Let us examine first the concept of testability. We test theories by facing them with basic statements, particular statements, that is, which are 'inter-subjectively verifiable', i.e. statements about the truth of which observers are agreed. (Popper uses the term 'statement' with the meaning which I have given to 'proposition', and I will follow his usage throughout this chapter.) Popper equates testability or falsifiability with content. For Popper the content of a theory T is measured by its 'empirical content' (the class of basic statements which T forbids) or its 'logical content' (the class of basic statements which are deducible from T).

How are we to compare the falsifiability of theories? Popper describes two methods – the subclass relation and dimensions. Let us consider first the subclass relation. Popper writes: 'A statement x is said to be falsifiable in a higher degree or better testable than a statement y . . . if and only if the class of potential falsifiers of x includes the class of potential falsifiers of y as a proper subclass' ([3] p. 115). Hence we ought to accept a theory T_1 rather than a theory T_2 if T_1 entails T_2. T_1 will entail T_2 if T_1 has greater universality (i.e. says the same thing about more objects) or T_1 has greater precision (i.e. says more about the same objects). We ought to prefer the more universal 'all orbits of heavenly bodies are ellipses' to 'all orbits of planets are ellipses', and the more precise 'all orbits of heavenly bodies are circles' to 'all orbits of heavenly bodies are ellipses'. The meaning of this criterion and the sense in which it is a criterion of falsifiability seem clear.

The second criterion of falsifiability – that of dimensions – is somewhat more problematical. To apply it we have to be able to identify basic statements of the same relative basicness or atomicity, statements which say the same amount about the world. The example which Popper gives is of two statements each ascribing a precise spatial location to a planet. They are equally atomic statements. Given that we can identify such statements, then a theory is n-dimensional, relative to a certain kind of atomic statement, if it needs a conjunction of $n + 1$ such atomic statements to falsify it. A theory T_1 is to be preferred to a theory T_2 if, relative to some kind of atomic statement, T_1 has m dimensions, T_2 has n dimensions, and $m < n$. (One possible difficulty with this definition is

that T_1 might have greater dimensions than T_2 relative to one kind of
atomic statement, but T_2 have greater dimensions than T_1 relative to
another kind of atomic statement. I can however think of no plausible
examples of this difficulty which must remain a mere theoretical one. A
possible solution, which might well be acceptable to Popper, would be
to say that in such cases the theory which has fewer dimensions relative
to the more atomic kind of atomic statement is to be preferred.)

By this criterion T_A, 'all planets move in circles', is to be preferred to
T_B, 'all planets move in parabolas'. For to falsify T_A you need a statement of the form 'there is an x such that x is a planet, and at some time
x is at P_1, and at another time x is at P_2, and at another time x is at P_3,
and at another time x is at P_4', when $P_1 \ldots P_4$ each uniquely identify a
place. Yet to falsify T_B you need a statement of the same form with 'and
at another time x is at P_5' added. If we suppose that each report of a
planetary position is a basic statement, then T_A is three-dimensional, and
T_B is four-dimensional. Again, to take a different example (and one which
is not Popper's) T_C, 'all ravens are black', is to be preferred to T_D, 'all
ravens except in Siberia are black, and all ravens in Siberia are not black'.
For to falsify T_C you need a statement of the form 'there is an x such
that x is a raven and x is not black', but to falsify T_D you need a statement either of the form 'there is an x such that x is a raven and x is not
black and x is not in Siberia' or of the form 'there is an x such that x is
a raven and x is black and x is in Siberia'. Given the reasonable assumption that 'x is black' said of some individual is not a more basic statement
than (i.e. tells us no less about the world than) 'x is not black', more basic-
statements compose either conjunction able to refute T_D than compose
the conjunction needed to refute T_C. For to refute T_C you need a statement 'there is an x such than x is a raven' conjoined with 'x is not black'.
To refute T_D you need as well 'x is not in Siberia'; or, instead of 'x is not
black' you need the no more basic statement 'x is black' conjoined with
'x is in Siberia'. If we suppose that statements ascribing to individuals
properties of being a raven, being black or not black, being in Siberia or
not in Siberia, are all equally basic, then, relative to basic statements of
this type, a conjunction of two basic statements is needed to refute T_C,
but of three basic statements to refute T_D. We will henceforward, to
simplify the argument, assume that all such statements are equally basic.

Now, given that we can identify statements of the same basicness, as
no doubt one often can intuitively, there is no doubt about the applicability of this test of falsifiability. What is more doubtful is whether falsifiability as so measured can always be equated with content. A theory,

according to Popper, has greater content than another theory, in Popper's sense of 'empirical content', in so far as it rules out 'a larger class of basic statements', and Popper seems to equate this with ruling out 'logically possible worlds' ([3] p. 113). Now a logically possible world is described by conjoining the existential statements which exhaustively describe that world. In a world in which there are ravens, such statements will describe the position as well as the colour of such ravens. T_C will therefore rule out worlds into the description of which enter either 'there is an x such that x is a raven and x is not black and x is in Siberia' or 'there is an x such that x is a raven and x is not black and x is not in Siberia'. T_D will rule out worlds into the description of which enter either 'there is an x such that x is a raven and x is not black and x is not in Siberia' or 'there is an x such that x is a raven and x is black and x is in Siberia'. Since all these statements are equally basic, the same number of worlds will be compatible with each. T_C and T_D both rule out worlds compatible with either of two conjunctions of three equally basic statements, and hence they rule out the same number of worlds. Hence two theories may rule out the same number of possible worlds, although one of them is more falsifiable. If anyone denies this conclusion because he denies that the components of the conjunctions cited are equally basic statements, the same conclusion can be reached by taking purely formal examples.[1] Let '$(\exists x)(\phi x)$', '$(\exists x)(\sim\phi x)$', '$(\exists x)(\psi x)$', and '$(\exists x)(\sim\psi x)$' be all equally basic statements. Let T_E rule out '$(\exists x)(\phi x)$' and so '$(\exists x)(\phi x . \psi x)$' and '$(\exists x)(\phi x . \sim\psi x)$'. Let T_F rule out '$(\exists x)(\phi x . \psi x)$' and '$(\exists x)(\sim\phi x . \sim\psi x)$'. T_E will be more falsifiable than T_F, for one basic statement will refute T_E while two are needed to refute T_F. Yet both T_E and T_F rule out the same number of conjunctions of two or more basic statements, and so the same number of possible worlds. Since it seems reasonable to understand by a theory having greater content than another theory that it rules out more possible worlds, content is not to be equated with falsifiability. Popper's dimensions criterion selects the more falsifiable theory, but not necessarily the theory of greater content.

At the end of his discussion of the 'dimensions' criterion, Popper adds a further criterion of choice between theories which he seems to regard as a further point about the 'dimensions' criterion. A theory T_A, 'all planets move in circles', has the same number of dimensions as a theory T_G, 'all planets move in ellipses through points P_1, P_2', where P_1 and P_2 each uniquely identify a place. A conjunction of four reports of planetary

(1) '$\exists$' is known as the existential quantifier. It governs the following bracket. '$(\exists x)(\phi x)$' means 'there is an x such that x is ϕ'; '$(\exists x)(\phi x \supset \psi x)$' means 'there is an x such that '$\phi x \supset \psi x$' is true of it'; and so on.

position is needed to refute each. However Popper feels that we ought to prefer T_A to T_G and he therefore proposes ([3] pp. 132-5) that, at any rate if we are considering theories about the motions of bodies, we ought to compare not only their dimensions, but also what he calls their generality. A theory is more general on Popper's definition, the larger the group of co-ordinate transformations with respect to which it is invariant. A consequence of this definition is that a theory of n dimensions which specifies a point through which a curve passes will be less general than a theory of the same number of dimensions which does not specify this, but merely specifies the shape of the curve. Hence T_G, although like T_A three-dimensional, is less general than T_A. Popper does not state explicitly a general rule for taking generality into account, but he seems to suggest that among theories of equal dimensions, we ought to prefer the more general. He does not however explain how, if at all, the rule that we should do so follows from his principle of preferring the more falsifiable theory. It is however fairly clear that the principle of generality does not follow from the principle of falsifiability. Both T_A and T_G above attribute equally specific properties (moving along a certain path) to the same objects (all planets), and the same number of observations would be needed to show each false, observations which are equally easy to obtain. T_A and T_G are equally falsifiable, yet differ in generality.

So much for falsifiability. Now for corroboration. Popper's general views on this are contained in Chapter X of *The Logic of Scientific Discovery*. On Popper's account theories with great falsifiability are submitted to test. We try hard to falsify them. In so far as we fail they are corroborated. Yet falsification means that a theory ceases to be corroborated. 'In general we regard inter-subjectively testable falsification as final (provided it is well tested) . . . A corroborative approval made at a later date . . . can replace a positive degree of corroboration by a negative one, but not *vice versa.*' Yet the more severe the tests to which a theory is subjected the better corroborated it is if it survives those tests.

In New Appendix ix of *The Logic of Scientific Discovery* Popper set forward a numerical measure of corroboration, of the severity of tests which a theory has passed. This was as follows. (I use the term 'corroboration' instead of the term actually used, 'confirmation', because Popper subsequently came to prefer the former term and because I am using 'confirmation' in the sense defined in Chapter I.) The corroboration of hypothesis h by evidence e is:

$$c(h/e) = \frac{P(e/h) - P(e)}{P(e/h) + P(e)} \{1 + (P(h) \times P(h/e))\}$$

where for all q and r $P(q)$ denotes the 'logical probability' in Popper's sense of q and $P(q/r)$ denotes the 'logical probability' of q, given that r is true. (I shall use '$P(\ /\ )$' and '$P(\ \ )$' with these meanings for the next three pages.) Popper suggested also a simpler alternative measure:

$$c(h/e) = \frac{P(e/h) - P(e)}{P(e/h) - P(e \cdot h) + P(e)}$$

For Popper for a nomological hypothesis h the logical probability $P(h)$ = 0 and also $P(h/e) = 0$ (given that, as assumed, e consists of 'basic statements') and so $P(e \cdot h) = 0$. Hence for such hypotheses both measures reduced to

$$c(h/e) = \frac{P(e/h) - P(e)}{P(e/h) + P(e)}$$

If h entails $\sim e$, i.e. the hypothesis is falsified,

$$c(h/e) = -1.$$

If h entails e, i.e. the evidence is solely statements predicted by the theory, e.g. that the relationships claimed by h are not violated at various spatio-temporal instants,

$$c(h/e) = \frac{1 - P(e)}{1 + P(e)}$$

That is, $c(h/e)$ depends solely on $P(e)$. For low $P(e)$, i.e. $P(e) \rightarrow 0$, $c(h/e)$ $\rightarrow 1$. For high $P(e)$, i.e. $P(e) \rightarrow 1$, $c(h/e) \rightarrow 0$. $P(e)$ is for Popper, as stressed earlier, a measure of the lack of content of e. Hence if e is a lot of evidence, $c(h/e) \rightarrow 1$; if e is little evidence $c(h/e) \rightarrow 0$.

The formal account of Popper's corroboration function which I have set forward is however misleading unless we take into account a crucial restriction which Popper placed on the content of e. He did not want $c(h/e)$ to be high just because the quantity of 'evidence in favour' of h was high. He wanted $c(h/e)$ to be high only if h had survived severe tests. He wrote: 'Our $c(h/e)$ can be adequately interpreted as degree of corroboration of h in the light of tests . . . only if e consists of reports of the outcome of sincere attempts to refute h, rather than attempts to verify h' ([3] p. 414). But then the definition becomes a highly subjective psychological one. Corroboration depends on the intentions of the scientist conducting the test. Does Popper really want to say that? Intuitively we are inclined to say that the acceptability of a theory is a matter of what it says and of the results of experiments, quite independently of the intentions of the scientists who did the experiments. A further question about the Popperian definition is how are we in detail to calculate $P(e)$? Given that $P(e) \to 0$ as the content of e increases, for what conjunction of how many statements of what degree of basicness, does $P(e) = 1/2$? Popper answers that he is concerned only with laying down the principles for comparing hypotheses in respect of corroboration, not with laying down rules for calculating absolute values of corroboration, and that to be able to order hypotheses in respect of corroboration is all that matters (see [4] p. 397).[1]

Lakatos has proposed that we replace $P(e)$ by $P(e/h')$ where h' is a 'rival touchstone theory'. h' must be 'a genuine rival theory with scientific interest' ([6] p. 415). One difficulty here is that if both h and h' are universal hypotheses which predict e, $P(e/h') = 1$ and so $c(h/e) = 0$. A universal hypothesis is then, on Lakatos's account, never corroborated by a test if a rival universal hypothesis predicts precisely the same test result. But surely a theory is corroborated if some other theories do not predict that precise result, even if the 'genuine rival' does? Another difficulty is what are we to say if there is no one genuine rival or more than one? Further, like Popper's, Lakatos's criterion is highly subjective. Corroboration here depends on what theories are the subject of scientists' current 'interest'. Yet further, scientists may be interested in theories on various grounds. With increasing interest in the history of science scientists may be interested in some theory h' on antiquarian grounds. But in that case surely a theory h is not corroborated by e if $P(e/h')$ is small. Clearly Lakatos would not count such an ancient theory as a 'genuine rival'. But in that case he must define what he means by 'a genuine rival theory with scientific interest'. One obvious thing that he might mean by this phrase is 'rival theory rendered probable, i.e. likely to be true, by the evidence'. But then corroboration could not be measured until probability (in the sense of likelihood of truth) had been measured. This alternative is hardly available to a strict Popperian, since in Popper's view – see my p. 211 – all theories not falsified by the evidence are equally likely

(1) Lakatos has an argument to show that $P(e)$ is always zero. He writes that 'the *absolute* rational betting quotient on any proposition is always zero' ([6] p. 407) and he supports this by claiming that 'all empirical propositions are universal because of the universal names inevitably recurring in them' ([6] p. 407). But a claim that 'all empirical propositions are universal' like a claim that all properties are dispositional or that all phenomena are mental, or all judgements subjective, clearly smudges distinctions. The occurrence in it of a term which *could* apply to an infinite number of objects hardly makes a proposition universal.

to be true. All rivals, on this view, are 'genuine rivals'. Lakatos needs to clarify what he understands by 'genuine rival theory with scientific interest'.

As we have noted, Popper connects his two criteria, testability (or falsifiability) and corroboration, by terming testability corroborability. For, he reasonably argues, in so far as a theory can be tested, it can survive tests and so be corroborated. But, more precisely, Popper equates the 'confirmability' i.e. corroborability of x with 'the maximum degree of confirmation' i.e. corroboration 'which a statement x can reach' ([3] p. 399). Does his corroboration formula allow him to do this? Lakatos points out ([6] pp. 410 and 414) that if $P(e)$ means simply the logical probability of e, then with increasing e, $P(e) \rightarrow 0$, and so $c(h/e) \rightarrow 1$ for all universal (in our terms, nomological) hypotheses, and hence all such hypotheses will have the same corroborability, viz. 1. Yet on the account given in Chapter 5 of *The Logic of Scientific Discovery* the testability of different universal hypotheses differs. A further discrepancy is that if e is to include only 'sincere' attempts to falsify h, corroborability will depend on the intentions of the investigator. On Lakatos's reinterpretation of Popper, corroborability will depend on the availability of rival theories, an unpredictable matter. On neither of these interpretations does corroborability possess the (often) objectively measurable character of the testability described by Popper in Chapter 5 of [3].

So much for the account given in *The Logic of Scientific Discovery* of the criteria of acceptability of scientific theories. I have been concerned to show that they are not quite as clear, nor do they form such a unity, as might at first sight appear. The next question is whether Popper's criteria are the right criteria for acceptability in his sense of acceptability.

Popper does not face up to the need to justify his criteria in *The Logic of Scientific Discovery*, but he does attempt a justification in *Conjectures and Refutations* [4], and this we must now consider.

The aim of science, Popper writes in *Conjectures and Refutations*, is comprehensive truth. The scientist seeks theories of as high verisimilitude as he can obtain. The verisimilitude of a theory is its closeness to the truth. It is a quantity which increases with its truth-content and decreases with its falsity-content. The truth-content of a theory is the class of true consequences of the theory, its falsity-content the class of false consequences of the theory. Science cannot ever reach comprehensive truth, but it can approach it. We can get theories which have greater verisimilitude than other theories, and the point of 'accepting' theories is to obtain the former.

Now the suggestion that science aims at high verisimilitude, in the sense of theories which have a lot of truth in them, even if some falsity as well, is a very reasonable one. According to this suggestion a completely true theory of small content is desirable, but even more desirable is a theory of large content with only a small amount of falsity in it. It is well known that scientists often live and work with a theory which they know to be slightly in error, if they cannot get a better theory; and the view that scientists seek theories of as high verisimilitude as they can get, explains their doing this. However, what is true of many scientists is not true of all. Scientists who are interested in theories solely in order to make predictions with them naturally seek high verisimilitude, and they no doubt are the majority of scientists. But scientists who seek to find the true laws of nature have no interest in theories in which there is a grain of falsity. Their aim is truth, not verisimilitude.

However, if we assume with Popper that the aim of science is verisimilitude, the question arises as to how it is to be measured. What counts as a theory of greater verisimilitude than another? The verisimilitude of a theory is a quantity which increases with the truth-content, the size of the class of true consequences of the theory, and decreases with the falsity-content, the size of the class of false consequences of the theory. Popper suggests that it is perfectly adequate for comparative purposes to measure verisimilitude by the truth-content minus the falsity-content (see [4] pp. 234 and 396). But how is the size of the class of true consequences or the size of the class of false consequences to be measured? We cannot measure these by the number of consequences in each, since all theories have an infinite number of true consequences, and a false theory will have an infinite number of false consequences as well, and mathematics ascribes no value to 'infinity minus infinity'.

This claim about the infinity of consequences can be shown as follows. If T is a true theory, and p a consequence of it and so a true proposition, so also is '$p \vee q$' where q is any proposition whatsoever. If a theory predicts that the temperature of a certain substance is 100°C, it predicts that it is either 100°C or lies between 100°C and 100·5°C inclusive, is either 100°C or between 100°C and 101°C inclusive, and so on. If T is a false theory, it will have at least one false consequence, which we will call 'p'. Hence it will have as false consequences all propositions of the form '$p \vee q$' where 'q' is any false proposition, the number of which is presumably infinite. It will also have as true consequences all propositions of the form '$p \vee q$', where 'q' is any true proposition, the number of which is presumably also infinite.

If we cannot for the reason given measure the size of the class of true consequences or the size of the class of false consequences by the number of propositions in each, how are we to measure these sizes? There seems no obvious way of doing this, and so of making the notion of verisimilitude precise enough to operate with. We do not have a precise concept of the size of a class of an infinite number of propositions. We would need to formulate one and the decision involved in doing so to use one way of measuring the size of a class of propositions rather than another would be an arbitrary one.

It is true that, even if we do not have quantitative values of the truth- and falsity-contents of theories, we may on occasion be able to say that the verisimilitude of one theory is greater than that of another, even if we cannot measure these values quantitatively. Indeed Popper regards 'the comparative use of the idea' of degree of verisimilitude as its 'main point' ([4] p. 234). However I can think only of one kind of case where we can compare the verisimilitude of theories. Consider two theories T_1 and T_2 related by the subclass relation, T_2 being a deductive consequence of T_1 but T_1 not a deductive consequence of T_2, T_1 and so T_2 being true theories. Then any consequence of T_2 will also be a consequence of T_1, but T_1 will have other consequences as well. Neither theory has any falsity-content, while T_1 has all the true consequences of T_2 and more as well. Hence the truth-content and so the verisimilitude of T_1 exceeds that of T_2. However if T_1 and T_2 are related by the subclass relation, T_2 being a deductive consequence of T_1, but T_1 not a deductive consequence of T_2, and T_1 is false, we cannot draw any general conclusions about whether T_1 has more or less verisimilitude than T_2. In a paper [5] Popper showed that if we have two theories whose contents are comparable (because one theory includes the other), or two theories whose truth-contents are comparable (because the class of true consequences of one includes the class of true consequences of the other), then both their content and their truth-content are comparable, and 'the comparison of their contents will always produce the same ordering' of the two theories 'as the comparison of their truth-contents' ([5] p. 352). If T_1 has greater content than T_2, it will have greater truth-content, and vice versa. Despite all this however, the fact remains that even if T_1 has greater truth-content than T_2 it may also have greater falsity-content too, and the extra falsity-content could outweigh the extra truth-content and so T_1 have less verisimilitude than T_2. Neither greater content nor greater truth-content guarantees greater verisimilitude.

But let us suppose that we do have a way, given knowledge of which of the consequences of a theory are true and which false, of measuring its verisimilitude. Then the further difficulty arises that we cannot *know* what the verisimilitude of a scientific theory is, because we cannot *know* the truth or falsity of each of the consequences of the theory – given, as Popper assumes, that our evidence consists solely of particular propositions reporting past events. A scientific theory consists of nomological propositions, and nomological propositions have as consequences predictions in the form of particular propositions about the future as well as the past. 'Of physical necessity all A's are B' tells us not merely about past A's, but about all points of space and instants of time, including future instants, that there is at them no object which is both A and not-B. We know the truth or falsity of some such consequences for the past but can only infer with probability from such evidence the truth or falsity of predictions about the future and about past instants unstudied. Knowing only probability of the truth of most of the consequences of a theory, we can only at most establish the probability of different values of verisimilitude, no more.

Popper and his followers have not fully faced this fact. Popper, admitting that we cannot know the verisimilitude of a theory, goes on 'but I can examine my guess of its verisimilitude critically and if it withstands severe criticism, then this fact may be taken as a good critical reason in favour of it' ([4] p. 234). But why? Popper does not tell us. Again, Watkins, writing in defence of Popper, admits that the fact that h_2 has stood up to tests better than h_1 'does not *verify* the claim that h_2 is closer to the truth than h_1 . . . But . . . there are good reasons for preferring this claim both to the (perverse) claim that h_1 has more verisimilitude than h_2, and to the (implausible) claim that h_1 and h_2 have an exactly similar verisimilitude' ([6] p. 281). But what are these 'good reasons'? Watkins does not tell us.

On the basis of the evidence then the scientist can at most establish the probability of different values of the verisimilitude – and this only given some way of measuring truth-content against falsity-content. In such a case he will aim at what we may loosely term 'high probable verisimilitude'. There are various ways in which 'probable verisimilitude' might be calculated, but different definitions will not greatly affect the point to be made. Let us say that 'probable verisimilitude' means expected verisimilitude, that is expected truth-content minus expected falsity-content. The expected truth-content of a theory is the sum of the possible values of the truth-content weighted by the probability of the occurrence of each. Thus, to take a very simple example, if there were three possible values of the truth-content, $\frac{1}{2}$, $\frac{3}{4}$ and 1, and the probability of the first two was each $\frac{1}{4}$ and of the latter was $\frac{1}{2}$, the expected value of the truth-content would be $\{(\frac{1}{2} \times \frac{1}{4}) + (\frac{3}{4} \times \frac{1}{4}) + (1 \times \frac{1}{2}) = \frac{13}{16}\}$.

Now we know at most two relevant things about a theory – its actual success to date and its probable future success. Hence expected verisimilitude equals (known past truth-content minus known past falsity-content) plus (expected future truth-content minus expected future falsity-content). Now the first term may be ignored, because the examined content of the theory will be very small, if not infinitesimally small, compared with the unexamined content, on any reasonable way of measuring these. We will only have examined a small number of spatio-temporal instants and found that the predictions made by our theory hold there; yet the theory makes predictions about all other instants of time and points of space in the Universe. So expected verisimilitude will equal (approximately) expected future truth-content minus expected future falsity-content.

Let us compare the content of theories by supposing that each theory consists of some finite number of units of content; the greater the content of the theory the greater the number of units. Then the expected verisimilitude of a theory will equal (number of units of content of the theory) times (expected verisimilitude of an average unit). The latter value will be quite independent of how extensive the theory is, its content. It will depend solely on the probability of an average prediction of the theory being true. For the greater that is, the more true and the less false predictions is it probable that a theory will make within an area of given size. Let us call this value, the expected verisimilitude of an average unit of content of the theory, v. Now if v is positive, the greater content a theory has, the greater its expected verisimilitude. If v is negative, the greater content a theory has, the less its expected verisimilitude. If $v = 0$ its content has no effect on its expected verisimilitude. So high content is only a desirable feature of theories when $v > 0$, that is when the expected number of true predictions exceeds the expected number of false predictions per unit of content, which is to say that the probability of an average prediction of the theory being true exceeds $\frac{1}{2}$. How is v to be calculated? Let us look at the matter first on Popperian assumptions about inference to the future and then on normal assumptions.

Popper is a sceptic who has been deeply influenced by Hume's view that what has happened is no reliable guide to what will happen in future (see [3] pp. 29f.). In Popper's view the probability of a prediction being true is unaffected by what has happened in the past, and in particular by the past success of a theory. In that case the probability of an average prediction being true and so the value of v, the expected verisimilitude of a unit of content of the theory, will be determined by *a priori* considerations alone, and be the same for all theories. In what possible way the value is determined however I cannot see. *A priori* it seems to me more plausible to suppose that $v < 0$ than that $v > 0$, since it seems to me that an arbitrary proposition is more likely to be false than to be true when we have no evidence bearing on its truth or falsity.

However in practice most of us suppose that the probability of the prediction of a theory being true depends on what is known to have happened in the past and in particular on the known past success of the theory. This book seeks to draw out and set down the criteria of probability which we use in practice, and so we will take the latter supposition to be correct.

We have not developed in this book any results which provide an easy formula for calculating the probability of an average prediction of a theory being true. However the results of Chapter VIII indicate that the probability of a particular proposition q is greater, the greater the probability of theories which predict q, and smaller, the greater the probability of theories which predict $\sim q$. However the expected verisimilitude of an average unit of a theory is by no means the same as the probability of that theory (or even as the probability of a unit of content of the theory), nor does the one necessarily increase when the other does. We can see this by considering two theories of equal content, T_1 and T_2. To say that T_1 is more probable than T_2 is to say that it is more probable that all the predictions of T_1 conjoined are true than that all the predictions of T_2 conjoined are true. Yet this could be because, although the vast majority of predictions of T_2 are more probably true than are those of T_1, T_2 contained one or two highly improbable predictions (perhaps even ones known to be false), and hence was as a whole less probable than T_1. Yet the expected number of true predictions of T_2 might exceed that of T_1. If it did, then, since the content of the two theories is the same, the expected verisimilitude of T_2 would exceed that of T_1, and so also would the expected verisimilitude per unit of content of T_2 exceed that of T_1. But to produce a general formula for calculating the expected verisimilitude of a theory or a unit of content of a theory would clearly be a complicated matter.

A recent book, *The Implications of Induction* by L. Jonathan Cohen [7], in effect provides a method for comparing theories in respect of their expected verisimilitude per unit of content. Cohen describes his book as concerned with 'the logical syntax of inductive support' ([7]

p. 6). As I commented in Chapter VI, what Cohen means by 'inductive support' is not initially very clear and he *seems* to mean by 'support' what I have called 'epistemic probability'. Under this interpretation, as I showed in Chapter VI, Cohen's account fails. However a remark on p. 50 of his book and subsequent correspondence with Cohen have made clear to me the true significance of his book. Cohen wrote in correspondence that 'supports' in his sense means 'something like "shows to be close to the truth" '. But by 'the truth' of a theory Cohen seems to mean not 'the truth' in Popper's sense of the whole truth about the universe but the truth about the domain covered by the theory. That being so, Cohen is providing a measure of the expected verisimilitude of a unit of content of a theory.

Cohen's theory is worked out rigorously and elegantly. The main idea is that a theory gains 'support' in so far as it passes tests under varied circumstances, circumstances whose occurrence has been found to make a difference to the operation of similar hypotheses. Thus the hypothesis that a certain drug taken by an expectant mother has no influence on the foetus gains support in so far as it is shown to work for different species of animals at different stages of pregnancy. If the drug is shown in one of these tests to have an effect on the foetus, that diminishes but does not reduce to zero the support provided for the theory taken as a whole. (Cohen assumes in saying that certain circumstances make a difference to the operation of similar hypotheses, the existence of a system of classification, for classifying hypotheses, circumstances etc. as similar to each other. Cohen does not discuss the grounds for choice between alternative systems of classification. This issue is one which I have discussed in Chapters VII and XI). Under the suggested interpretation Cohen's system is highly plausible, and may well follow from the principles of epistemic probability which I have set out.

In so far then as scientists seek only high expected verisimilitude per unit of content of a theory, or theories the probability of an average prediction of which being true is high, they will seek theories with high 'inductive support' on some such measure as Cohen's. In so far as they seek theories with high expected verisimilitude, they will seek theories for which ((units of content) times (expected verisimilitude per unit of content)) is high. Let us consider first how theories with high expected verisimilitude per unit of content (v) are to be obtained. Theories must first be invented and then tested, in the sense that evidence must be sought which affects the probability of their predictions. Among theories which have been invented and put forward for our consideration which ought we to test? Clearly those which on present evidence have high v, and which are most likely to have v raised by testing. v will be raised by testing in so far as the theory up for tests passes those tests. In so far as the tests are severe, as Popper demands that they should be, and so are extensive tests of the weakest parts of the theory, that theory among others covering the same area which is most likely to pass such tests is that theory which is more probable than those other alternatives. For such a theory it is more probable that all its predictions, including its weakest ones, are true, than that the predictions of any alternative are true. Suppose now that we seek not high v, but high expected verisimilitude, that is high (v times (units of content)). For $v > 0$, as we saw earlier, this will be greater, the more units of content. But it will also be greater, the greater is v. Testing cannot affect the content of a theory, but it can affect the value of v, and, as we have seen, if we are to test theories for which v is most likely to be raised by testing, we ought to test theories which are most probable.

I conclude that the scientist who seeks theories with high expected verisimilitude (whether total or per unit of content of the theory) ought to 'accept' (in sense [C]), other things being equal, theories which are more probable than alternatives. Some of the 'other things', e.g. content, we have already noticed. But other relevant factors are such things as the ease or cheapness of testing. The result that the scientist seeking high expected verisimilitude (or high expected verisimilitude per unit of content) ought, other things being equal, to 'accept' the more probable hypothesis clearly holds generally, whether or not, as we have been assuming throughout this chapter, that hypothesis is a scientific theory. The argument above applies generally. The expected verisimilitude of a proposition will be raised if it passes tests and not if it does not. A proposition is more likely to pass tests, the more probable it is. Hence, other things being equal, we ought to test the more probable propositions.

If the scientist seeks not verisimilitude but truth in some area, he will be seeking the hypothesis for that area which has high probability. Which hypotheses ought he to test in order to reach theories of high probability? Other things being equal, those which are most likely to pass tests, that is, those which are already more probable than other hypotheses. I conclude that whether the scientist seeks truth or verisimilitude or high verisimilitude in some area, he ought, other things being equal, to submit to tests, that is accept in sense [C], those hypotheses which, before the tests, are more probable. We thus reach the same conclusion about acceptability in sense [C] as about acceptability in other senses.

Bibliography

For the Neyman-Pearson theory of testing, see:

[1] J. NEYMAN and E. S. PEARSON 'On the Problem of the Most Efficient Tests of Statistical Hypotheses' *Philosophical Translations of the Royal Society,* Series A, 1933, **231**, 289–337.

For commentary on statistical theories of testing, see:

[2] I. HACKING *The Logic of Statistical Inference,* Cambridge, 1965, Chapters 6 and 7.

For Popper's account of acceptability, see:

[3] KARL R. POPPER *The Logic of Scientific Discovery,* London, 1959.

[4] KARL R. POPPER *Conjectures and Refutations,* 3rd edition, London, 1969, especially Chapter X.

[5] KARL R. POPPER 'A Theorem on Truth Content' in Paul K. Feyerabend and Grover Maxwell (eds.) *Mind, Matter, and Method,* Minneapolis, 1966.

For commentary on Popper, see:

[6] Articles by J. W. N. WATKINS and I. LAKATOS in I. Lakatos (ed.) *The Problem of Inductive Logic,* Amsterdam, 1968, pp. 271–82 and pp. 316–417.

For Cohen's theory of inductive support, see:

[7] L. JONATHAN COHEN *The Implications of Induction,* London, 1970.

Index